SpringerBriefs in Digital Spaces

W0090565

Series Editor

Ahmed Bounfour

For further volumes:
http://www.springer.com/series/10461

Olivier Dupouet · Tatiana Bouzdine-Chameeva
C. Lakshman

Innovation from Information Systems

An Ambidexterity Approach

 Springer

Olivier Dupouet
BEM Bordeaux Management School
Talence
France

C. Lakshman
BEM Bordeaux Management School
Talence
France

Tatiana Bouzdine-Chameeva
BEM Bordeaux Management School
Talence
France

ISSN 2193-5890
ISSN 2193-5904   (electronic)
ISBN 978-3-642-32875-6
ISBN 978-3-642-32876-3   (eBook)
DOI 10.1007/978-3-642-32876-3
Springer Heidelberg New York Dordrecht London

Library of Congress Control Number: 2012945746

Printed on acid-free paper

Springer is part of Springer Science+Business Media (www.springer.com)

Contents

Innovation from Information Systems: An Ambidexterity Approach

1 Executive Summary

The overall innovation process at Legrand can be described as a cycle involving the entire organization, with the cycle itself consisting of an upward stream and a downward stream (cf. Andriopoulos and Lewis 2009). In the upward phase of this cycle, Ideas and concepts for future products are generated throughout the organization, including operational levels in the different countries where the group operates. These ideas are then sent to headquarters, to the different divisions, where ideas are gathered, further assessed and developed. Projects deemed most promising are then submitted to top management that decides on which projects to invest. Once this decision is made, innovation enters the downward phase of its cycle. This phase begins with an exploration phase in which, R&D on new products is conducted centrally in some R&D centers, mostly located in France and Italy. Once prototyped, the products are then produced, marketed and sold in the different subsidiaries, close to their markets.

Introducing time as an explanatory variable of a firm's ambidexterity, our case study proposes a process view on ambidexterity. To date, the bulk of the literature on ambidexterity presents static ambidextrous settings (Raisch et al. 2009). The focus has been largely on structural design that could enable ambidexterity in organizations (O'Reilly and Tushman 2004; Gibson and Birkinshaw 2004). Yet, several studies point to the importance of time in developing new products, processes or capabilities within the firm (Brown and Eisenhardt 1997; Nickerson and Zenger 2002; Siggelkow and Levinthal 2003; Westerman et al. 2006). In this regard, our key contribution and our main finding is that in order to understand the mechanisms through which MNCs manage ambidexterity, one must take a process view on ambidexterity. In the studied company, ambidexterity is achieved through a process in which the entire organization is involved. It starts with idea generation from any place in the organization. These ideas are funneled towards specific explorative teams that further develop and refine them. These explorative projects

O. Dupouet et al., *Innovation from Information Systems*,
SpringerBriefs in Digital Spaces, DOI: 10.1007/978-3-642-32876-3_1,
© The Author(s) 2013

are then submitted to the evaluation of the top management that manages the overall firm's technological portfolio. Agreed projects are then progressively transformed into exploitable products, following a stepwise transformative learning process.

Another interesting aspect unveiled by our study is the idea of multi-level ambidexterity, i.e., exploration at one level compensated by exploitation at another level and vice versa. At Legrand, our analysis reveals that the operational level, in subsidiaries is rather focused on exploitation, i.e. their primary driver is the achievement of production and sales objectives. In terms of time allocation, much more is spent on the realization of the prescribed tasks than on the generation of new ideas. At Middle management level, in particular at the headquarter, there is a much greater stress on exploration in that they are explicitly required to generate innovation (product innovation in divisions and process and managerial innovations in support functions). At the top management level, individuals pay great attention to the financial risk entailed by novel projects and can thus be characterized as being rather exploitative.

Consistent with prior empirical evidence, we also find evidence of the simultaneous presence of contextual and structural ambidexterity, depending on the stage of development (Raisch 2008; Raisch et al. 2009). During idea formation, the ambidexterity is mostly contextual (Gibson and Birkinshaw 2004). Explorative activities (i.e. ideas generation) are carried out in the course of daily activities, while being encouraged and supported by a favorable managerial context (i.e. proposing new ideas is encouraged and valued). As ideas gain strength and are progressively turned into R&D projects, the ambidexterity becomes increasingly structural (O'Reilly and Tushman 2004). People in charge of developing these new ideas are more specialized, organized in dedicated teams and clearly identified work structure. Andriopoulos and Lewis (2009) note the importance for the success of an innovative firm to mix contextual and structural ambidexterity.

In terms of information systems, we think that for idea generation, agents should be provided tools for exploration (such as electronic communities of practice) and tools for exploitation (such as enterprise information systems), but that the balance between the two kinds of activities should be decided by agents themselves.

In this balancing act, the upward stream of this process is made possible by a real option reasoning process. This process allows the generation of options that can be exercised or abandoned by the organization, thereby providing it with flexibility in the early phases of exploration. The downward stream of this process relies heavily on the matrix structure of the organization that permits both to transfer knowledge along divisions from one point to another in the organization, while also introducing further check points controlled by middle managers that can assess the value of an innovation further downstream in the process.

From a prospective standpoint, this opens up possibilities for equipping firms with specific information systems. The real option metaphor allows us to think that a decision support system, inspired from real options as used in the energy sector for instance, could be conceived based on this idea to help in balancing explorative and exploitive projects in the technologies portfolio of the firm. Such a decision support system could be used by the top management that make the investment decisions.

While the upward stream is a differentiating process by which explorative activities are progressively separated from exploitative ones, the downward stream follows the opposite direction. Once new products have been prototyped and tested, they are re-inserted in the normal processes of the organization. The downward stream is thus an integrating process by which new and old knowledge are combined (Raisch et al. 2009).

In this downward stream, information systems could be useful in supporting the transformative learning process. This process is made of two principal components: knowledge transfer and knowledge interpretation (Turner and Makhija 2006). Regarding the transfer aspect, an information system should help in translating knowledge into representations understandable by receivers (here, people from subsidiaries). Such a system should help in depicting and reflecting upon an understanding of the new knowledge arriving at the subsidiary. It would thus not be a simple transfer of codified knowledge but a tool that would help in transforming knowledge from one form into another. In addition, an information system serving this part of the process could also help in creating horizontal linkages between subsidiaries, thereby speeding up cross-fertilization and collective sense-making.

We provide, to our knowledge, the first empirical investigation and findings related to the theoretically derived HR configurations discussed in the literature. Recent work that has investigated the human resources architectures that facilitate ambidextrous learning (Kang and Snell 2009) has identified two alternative intellectual capital architectures that are well suited for ambidextrous organizational learning. The first of these two alternative architectures, labeled 'refined interpolation', depends on combining a specialist human capital, with cooperative social capital, and an organic organizational capital. The second architecture, labeled 'disciplined extrapolation' uses a combination of a generalist human capital, an entrepreneurial social capital, and a mechanistic organizational capital. Our findings reveal that the approach of Legrand is clearly one of disciplined extrapolation. This HR architecture, in conjunction with the matrix organizational design, aids both the upward and downward streams of the innovation process detailed by us. The human capital at Legrand is primarily generalist with a predominant use of expatriates in the MNC's subsidiaries, combined with other host country nationals for support. The social capital at Legrand is primarily entrepreneurial, with historically significant levels of autonomy to the different country subsidiaries, which is now being balanced with higher levels of standardization. Still, there remains a sufficient level of autonomy in the hands of the country heads to contribute to balancing exploration and exploitation at the global level. The organizational capital at Legrand is primarily mechanistic, with clearly defined procedures, sufficiently centralized, and systematically pursued by the corporate office in ensuring that innovation yields significant financial results.

2 Introduction

Organisational learning is critical for success of global companies in the complex and turbulent environment of the twenty-first century. In this regard, recent research emphasized the importance for firms to balance exploration, necessary to anticipate and prepare for the future, and exploitation, necessary to reap the immediate benefits of the current knowledge (March 1991). Specific designs are often seen as a means for firms to maximize the benefits of both exploration and exploitation. In particular, the concept of structural ambidexterity has received considerable attention in recent years (Tushman and O'Reilly 1996; Raisch et al. 2009). In such a setting, firms design and set specific units, including specific HR configurations, to encapsulate exploration and exploitation. In this way, firms are seen as being able to simultaneously carry out both activities.

The geographical dispersion of companies has called for the development and increased use of virtual organisations: nowadays, to a greater or lesser extent, information systems are integrated parts of organizational functioning, and firms must seek to smoothly articulate their design with the systems they use (DeSanctis and Monge 1999; Ahuja and Carley 1999; Jensen et al. 2009). Moreover, some research focused on the contribution of specific systems (including organisational, HR, and IS) to exploration or exploitation activities (Schulz 2001; Haas and Hansen 2007; Kane and Alavi 2007; Kang and Snell 2009).

Nonetheless, much remains to be understood in terms of what combination of organisational structures, the associated HR architectures, and information systems might be best suited for supporting ambidextrous design. In other words, although we have results regarding the links between organizational designs, information systems and either exploration or exploitation, we know little on how information systems can support the management of the tensions between the two.

Our project aims to contribute to addressing this issue. More precisely, exploration and exploitation activities on firms' learning and the evolution of their knowledge base (March 1991, 1996, 2006; Levinthal and March 1993) have increasingly been seen as important drivers of firms' dynamic capabilities (McGrath 2001; Nerkar 2003).

There is a growing number of works stressing the importance of differentiation and integration mechanisms for ambidexterity to become a dynamic capability (Raisch et al. 2009; Jansen et al. 2009; Andriopoulos and Lewis 2009; Westerman et al. 2006). Differentiating exploration and exploitation allows the pursuit of both, without one thwarting the other. In this way, explorative activities can be pursued far from the existing core knowledge base of the firm. Integrating back the two activities is then necessary to reap the value of exploration efforts and sets the ground for further exploration.

However, to date, the focus has been largely on static structural design that could enable differentiation and integration in organizations, investigating the structures and mechanisms that should exist to enable differentiation and integration, as well as their static arrangements (O'Reilly and Tushman 2008;

Jansen et al. 2009; Andriopoulos and Lewis 2009). Yet, organizational design, though essential, is only one part of the story. We also need to explain how differentiation and integration are actually carried out in organizations, and how these processes unfold over time (Siggelkow and Levinthal 2003; Westerman et al. 2006; Raisch et al. 2009). We contribute to this literature by introducing the time dimension in our understanding of differentiation and integration mechanisms. More specifically, we examine how differentiation and integration are used throughout a process to dynamically manage exploration and exploitation. We thus take a much needed organizational process perspective in examining ambidexterity initiatives in multinational corporations (MNC).

Further, once identified the different stages of processual ambidexterity and the different mechanisms at work in each, it will be possible to identify or propose information systems that would best support it.

We do so by resorting to a case study conducted in a MNC, which is a world leader in its sector of activity. Because there is no previous work on that precise topic, a single explorative case study seems an appropriate method. Further, we chose a MNC because by nature it relies on information systems to communicate and coordinate among geographically dispersed locations. Moreover, given its numerous lines of business and its industrial nature, Legrand has to manage exploration, in that it has to continuously introduce new products on its markets, and exploitation, in that it has to produce a great quantity of units in a standardized manner. The firm thus seems to be the appropriate place to observe exploration and exploitation, as well as a mature information infrastructure.

Our findings reveal a processual view on ambidexterity, characterized by a progressive differentiation of exploration and exploitation activities, followed by a progressive integration. We also find that this process is scattered throughout the firm and is realized and monitored by actors from operational, tactic and strategic levels. Along the process, different information systems are mobilized to strike the appropriate balance between exploration and exploitation.

The predominant factors we have extracted from the case study permit us to develop a formal model and emergent scenarios: we determine the key variables and processes explaining the interplay between organisational design (including HR architecture) and information systems in managing ambidexterity. We also formalise these findings in three new mathematical models which describe the different phases of the studied process: the impact of organisation structure; the continuous innovation process and the knowledge transition phase of ambidexterity.

The report is organized as follows. In the first part, we present the main theoretical concepts relevant to the topics we aim to address. In particular, we provide a definition of ambidexterity and the specificity of multinational corporations as contexts in which to deploy it. We then present our empirical results, drawn from a single case study carried out at Legrand, world leader in electric appliances. We then derive from the case theoretical implications and propositions, stressing in particular a process view of ambidexterity and likely conditions of success.

The following part presents a more formal model of what we observed and in the last section before concluding our report we focus on mathematical modelling for the functioning of innovation process as it emerges in a multi-national company.

3 Literature Review

3.1 Ambidexterity

3.1.1 Balancing Exploration and Exploitation and the Concept of Ambidexterity

Research in management has stressed the importance of exploration and exploitation activities on firms' learning and the evolution of their knowledge base (March 1991, 1996, 2006; Levinthal and March 1993) and these have increasingly been seen as important drivers of firms' dynamic capabilities (McGrath 2001; Nerkar 2003).

For March (1991), these two kinds of activities rely on different organizational learning logics: *"The essence of exploitation is the refinement and extension of existing competencies, technologies, and paradigms…The essence of exploration is experimentation with new alternatives"*. Exploration includes activities oriented towards variation, risk taking, flexibility, discovery, whereas exploitation refers to efficiency, selection and execution (March 1991). Burgelman (1991, 2002) distinguishes at the strategic level a dual process that stems from, respectively, the willingness to increase variance and the willingness to reduce it. For Baum et al. (2000) *"exploitation refers to learning gained* via *local search, experiential refinement, and selection and reuse of existing routines. Exploration refers to learning gained through processes of concerted variation, planned experimentation, and play"* (2000). In a similar way, for Benner and Tushman (2003), *"Exploitative innovations involve improvements in existing components and build on the existing technological trajectory, whereas exploratory innovation involves a shift to a different technological trajectory"*. For Gavetti and Levinthal (2000), who focus more on cognitive processes, exploration is oriented towards the future and relies on a revision of beliefs and agents' anticipations, whereas exploitation is oriented towards the past and relies on experience. In economic terms, incomes associated with exploration are more distant in time and uncertain, whereas revenues derived from exploitation are more immediate and certain. Exploration can thus lead to substantial variations in performance and lead to tremendous success, but also numerous failures, whereas exploitation is conducive of more steady performances (He and Wong 2004). In the R&D activities, exploration and exploitation are usually characterized respectively as local versus distant search (Sidhu et al. 2007; Rosenkoph and Nerkar 2001), or in terms of breadth versus depth of search (Katila and Ahuja 2002).

Literature on organizational learning shows that the way firms allocate their resources and attention to these two learning activities can be critical for their performances and the development of dynamic capabilities in an uncertain and changing environment. As March (1991) put it *"Adaptive systems that engage in exploration to the exclusion of exploitation are likely to find that they suffer the costs of experimentation without gaining many of the benefits. They exhibit too many undeveloped new ideas and too little distinctive competence. Conversely, systems that engage in exploitation to the exclusion of exploration are likely to find themselves trapped in suboptimal stable equilibria"* (1991). Organizations with the ability to orchestrate and balance the two activities in a dynamic way (exploration to create new opportunities for future revenues and exploitation to reap benefits of their past exploration efforts) are called ambidextrous and are viewed as the most apt to develop and maintain a competitive advantage (Gibson and Birkinshaw 2004; He and Wong 2004; Jansen et al. 2006; Sidhu et al. 2007).

Although the importance of ambidexterity is widely acknowledged, differences and tensions existing between the two kinds of learning lead many authors to stress their incompatible nature (in terms notably of culture and organizational routines) and to explicit difficulties raised in balancing them (March 1991; Levinthal and March 1993; McGrath 1999). As, for instance, Levinthal and March (1993) note, *"where situations or proper responses are numerous and shifting, it is harder to specify and realize optimal inventories of knowledge. By the time knowledge is needed, it is too late to gain it; before knowledge is needed, it is hard to specify precisely what knowledge might be required or useful. It is necessary to create inventories of competencies that might be used later without knowing precisely what future demands will be"*.

Over time, different organizational designs and mechanisms for ambidexterity have been identified. As Simsek et al. (2009) note firms may choose either to conduct exploration and exploitation activities simultaneously or in a sequential manner, and can choose to conduct them in distinct organizational entities or in the same entities, asking individuals to bear the burden of ambidexterity. In the next section, we present in turn these different mechanisms.

3.1.2 Different Forms of Ambidexterity

Punctuated Equilibrium

In his seminal 1991 article, March contends that exploration and exploitation are essentially antagonistic and cannot be pursued simultaneously by a firm. Because the two processes compete for scarce resources and because of their inherently different nature, March argues, exploration and exploitation are mutually exclusive. As Gupta et al. (2006) note, an organizational answer to this trade-off is to separate temporally the two kinds of activities. To cope with the tensions between the two, firms can alternate periods of exploitation and exploration. This cyclical process has been termed punctuated equilibrium.

According to the punctuated equilibrium model (Ford and Ford 1994; Gersick 1991; Tushman and Romanelli 1985) firms spend most of their life in exploitation phases. However, at certain points in their history, discrepancies between their environments and internal processes, procedures, and knowledge become so great that firms trigger major changes and undergo complete reorganization and strategic reorientation (Ohja et al. 1997). Once changes have been implemented, firms stabilize the new processes and activities, and enter a new stable exploitation phase. The key mechanism at the heart of punctuated equilibrium is thus an unfreeze-modify-refreeze sequence (Weick and Quinn 1999; Burgelman 2002).

Such a roll-over does not go without risk. Indeed, as Raisch (2008) accounts, many companies, such as ABB, HP or Sony, failed, or at least underwent great difficulties in conducting such radical changes. Risks of failure in these processes stem from disturbance felt by people, the sometimes deep modifications of hierarchical relationships, responsibilities linkages and chains of command. In this respect, Sastry (1997) finds out that successful change requires a period of trial during which performance assessment is suspended after modifications have been made. Simsek et al. (2009) adds that *ad hoc* mechanisms to help conflict resolution and encourage open communications between managers should be set to ease the transition from one equilibrium state to another.

Adopting a punctuated equilibrium approach is also extremely costly. As an example, Raisch (2008) notes that such a change at Deutsche Bank entailed direct costs worth more than € 1 billion. To these costs, must be added losses in productivity due to resistance to change and perturbations in clients' relationships (Westerman et al. 2006).

For these reasons, punctuated equilibrium must be used cautiously by firms, and in specific situations. The decision to resort to this approach can be driven by environmental changes that cause decline in performance. For instance major technological changes to which the firm is forced to adapt (e.g. IBM moving from mechanical machines to computers (Taylor and Helfat 2009) or Intel moving from hard drive to microprocessor core activitiy (Burgelman 2002)) are good examples of such environmental changes. Other examples could be major reorganizations (e.g. BMW moving from a centralized to a decentralized structure (Raisch 2008) or NUMMI switching from mass to lean production (Adler et al. 1999) decided in the face of severe organizational inefficiencies.

Thus, punctuated equilibrium is advisable to conduct major changes that affect the entire organization. It is a traumatic, costly, and risky process that can only be implemented when there are no other choices left (Raisch 2008). However, firms continuously face situations in which exploration and exploitation needs to be balanced, while not willing to modify their structures. Moreover, arguably, if firms are able to smoothly balance exploration and exploitation in the course of their activities, they may not face the need to conduct punctuated equilibrium-like changes (Benner and Tushman 2003). To smoothly manage the trade-off between exploration and exploitation, firms have to opt for specific organizational designs that permit them to continuously arbitrate between the two kinds of activities.

Recent years have witnessed a surge in the literature around the concept of organizational ambidexterity, that is firms' ability to simultaneously conduct exploration and exploitation (Gupta et al. 2006; Raisch and Birkinshaw 2008). Two approaches have been proposed to account for the means through which firms can be ambidextrous: structural and contextual ambidexterity. Within the structural ambidexterity framework (Benner and Tushman 2003; O'Reilly and Tushman 2004; Gilbert 2006), agile firms are defined as having developed specific entities dedicated exclusively either to exploration or exploitation. Proponents of the second kind of ambidexterity, contextual ambidexterity, adopt a different standpoint. They believe ambidexterity is achieved by individuals, in a context framed by management that promotes both knowledge creativity and efficiency (Gibson and Birkinshaw 2004; Kang and Snell 2009).

Structural Ambidexterity

Structural ambidexterity (Benner and Tushman 2003; O'Reilly and Tushman 2004) proposes that agile firms have developed specific entities or structures dedicated to either exploration or exploitation. Depending on the orientation of a given structure, specific incentive systems, controls and governance mechanisms should be attached to it. The firm's overall coherence and coordination are ensured by top management which possesses the necessary skills to arbitrate between exploration and exploitation and to decide the orientation of future exploration efforts.

For most authors (Gibson and Birkinshaw 2004; Benner and Tushman 2003; Burgelman 2002), organisational ambidexterity is presented as an alternative to punctuated equilibrium models, that suppose sequential shifts between exploration and exploitation (Gersick 1991). By contrast, the notion of ambidexterity captures the idea that the two logics can be simultaneously supported in one single organisation. There exist at the moment two main models around which the debate on how exploration and exploitation can coexist: structural ambidexterity and contextual ambidexterity. These two approaches differ in their understanding of structures and governance mechanisms that supports and steers the two logics.

Within the framework of structural ambidexterity (Benner and Tushman 2003; O'Reilly and Tushman 2004; Gilbert 2005), ambidextrous firms have developed entities dedicated to either exploration or exploitation. Depending on the kind of objective pursued by the unit, specific incentive, control and governance systems are attached to it. Coordination and global coherence of the firm is ensured by top managers who are seen as the only actors endowed with competencies and abilities necessary to decide future orientations of the firm.

This view of ambidexterity takes well into account the strategic and managerial dimensions necessary to carry out innovations while pursuing established activities (Mirow et al. 2007), by placing the top management at the heart of the mechanism. Besides, it is congruent with numerous studies underlying the necessity to structurally separate exploration, better supported by organic structures, and exploitation, that requires more bureaucratic structures (Adler et al. 1999; Stieglitz and Heine 2007). However, structural ambidexterity leaves two blind spots. First, it remains silent on the way the knowledge base upon which decisions to explore/exploit is constituted. In structural ambidexterity, the whole process relies on the top management borrowing the traits of the Schumpeterian entrepreneur able to scan her environment and to detect the future direction of development. Second, since only the top managers' abilities guarantee the overall cohesion of the firm (O'Reilly and Tushman 2004; Raisch and Birkinshaw 2008), the structural approach poses the problem of the integration of the various activities within the firm.

Contextual Ambidexterity

According to Gibson and Birkinshaw, who forged the concept, contextual ambidexterity is *"the ability to realize simultaneously flexibility and alignment throughout an entire business unit"* (Gibson and Birkinshaw 2004, p. 210). This assertion relies on a specific vision of exploration and exploitation processes that, according to us, can be characterized along three dimensions: the cognitive distance between knowledge held and knowledge to be acquired, exploration and exploitation viewed as two ends of a continuum and a specific risk management process.

The distinction between exploration and exploitation takes a great variety of shapes and has been applied in a great number of contexts (Li et al. 2008). Nonetheless, all authors agree that the distinction between exploration and exploitation can be defined in terms of distance between knowledge held by the considered agent and those (s)he seeks to acquire (March 1991; Gupta et al. 2006; Li et al. 2008). Hence, even if spaces and metrics vary (Li et al. 2008), it is generally admitted that exploration consists of fetching knowledge remote from the ones already possessed, whereas exploitation is a local search around the mastered corpus of knowledge (Gupta et al. 2006).

Exploration and exploitation can also be defined either as orthogonal or as two extremes of the same continuum (Gupta et al. 2006). The central proposition of contextual ambidexterity, simultaneously realize alignment and flexibility in the same value chain, supposes a permanent link between differentiation and integration of new and old knowledge. This can only be possible if exploration and exploitation are considered not as substantially different but as two expressions of the same phenomenon (Raisch et al. 2009).

In his original proposition, March (1991) underlines that exploration distinguishes itself from exploitation by entailing greater risk. Researchers working on structural ambidexterity observe enterprises at a specific moment of their history, when hierarchy has decided to significantly modify their organization's portfolio of competencies and/or activities. In such a situation, investments are important, risk is high, but concern a single alternative. On the contrary, the idea that contextual ambidexterity allows a continuous balance between exploration and exploitation implies that risk taking is more frequent, but concerns smaller investments distributed over a great number of actors. Exploration and exploitation in that case concern mostly knowledge and concepts and do not necessitate costly realizations. Let us note that these two modes of risk management are not mutually exclusive. Gibson and Birkinshaw (2004) note in this respect that the outputs of exploration and exploitation endeavors carried out in a contextual way can then be brought to the top management level in order to make investment decisions.

Based on this representation of exploration and exploitation mechanisms, Gibson and Birkinshaw (2004) propose to operationalize contextual ambidexterity by placing individuals at the center. At the heart of Gibson and Birkinshaw's (2004) proposal is the idea that tensions between exploration and exploitation are solved at the individual level, whatever their positions in the organizational chart (Gibson and Birkinshaw 2004; Taylor and Helfat 2009; Mom et al. 2007). Contextual ambidexterity is operationalized in a situated manner, in the course of agents' daily activities. Ambidextrous individuals are endowed with the ability to conduct simultaneously, exploitation activities, producing knowledge actionable in their functions, and exploration activities, that imply active engagement in new knowledge and activities. Contextual ambidexterity is facilitated by a dense personal network (Mom et al. 2007), and by the realization of inter-functional coordination tasks (Taylor and Helfat 2009; Mom et al. 2007).

Nonetheless, in order to give the opportunity to individuals to be ambidextrous, managers must create a favorable environment. Hence, the second key proposal of contextual ambidexterity is that, under the condition that the hierarchy provides an organizational setting combining incentives for efficiency and creativity, firm's members will organize their work time so as to combine exploration of new knowledge and exploitation of already mastered knowledge. Relying on the work by Ghoshal and Bartlett (1994), Gibson and Birkinshaw (2004) show that performance is insured by setting ambitious objectives (*stretch*) and rigorous control systems (*discipline*). A social context that favors knowledge creation and sharing must rely on assistance and support mechanisms (*support*) and the development of a culture of sharing and trust (*trust*). The hierarchy is also in charge of recognizing new knowledge and practices, to assess and stabilize and institutionalize those that seem the most valuable.

Setting contextual ambidexterity supposes the existence of a system that underlies and operationalizes the context in which individuals are free to choose how to balance exploration and exploitation. In this respect, existing contributions are scarce and remains more descriptive than explanatory (Brion et al. 2008).

In what concerns mathematical modeling of exploration and exploitation activities in the firms, a growing stream of research focused on the models of joint innovation introducing a measure of distance in knowledge space which determines the effectiveness of collaboration (see for example, Nooteboom 2000; Cowan et al. 2007; Peretto and Smulders 2002; Mowery et al. 1998; Bouzdine-Chameeva and Dupouët 2008). Feng et al. (2010) suggested modeling the different types of organization structure: nearly isolated subgroups; semi-isolated sub groups and random network and construct a generalized payoff function to examine the impact of inter-relational network structure on organizational-level performance. These studies reveal a general tendency for a simulation approach.

3.2 The Use of Information Systems for Sustaining Ambidexterity

3.2.1 Virtual Organizations

Because of their geographical dispersion, MNCs usually have to resort intensively to information and communication technologies to communicate and coordinate (Magnusson 2004). In this respect, MNCs display features of virtual organizations. DeSanctis and Monge, in their seminal article, define virtual organizations as: "*A collection of geographically distributed, functionally and/or culturally diverse entities that are linked by electronic forms of communication and rely on lateral, dynamic relationships for coordination. Despite its diffuse nature, a common identity holds the organization together in the minds of members, customers, or other constituents.*" (1999, p. 693). As Jensen et al. (2009) note, following this definition, MNCs can display more or less great virtuality. Virtuality then becomes a dimension of the firm's organizational design with which management can play to achieve efficiency: information systems are integrated parts of organizational functioning, and firms must seek to smoothly articulate their design with the systems they use (De Sanctis and Monge 1999; Ahuja and Carley 1999; Jensen et al. 2009).

More closely related to the topic of this study, some research focused on the contribution of specific information systems to exploration or exploitation activities (Schulz 2001; Haas and Hansen 2007; Kane and Alavi 2007).

First, information technologies can be used for knowledge circulation among geographically dispersed entities. Research in this area reveals in particular that there is no substitutability between different communications means (Hansen 1999; Schulz 2001; Haas and Hansen 2007). Haas and Hansen (2007), for instance, show that resorting to face-to-face communications contribute to increasing the quality of work, while resorting to electronic documents exchange increase work speed, but that no communication means fulfil the two requirements. Hence, the choice of communication means should be made to serve specific ends (Hansen et al. 1999). This is congruent with numerous works on social networks and the type of ties that should be set in order to transmit specific type of knowledge (Hansen 1999). This is also true in defining the directionalities of the communication channels, as discussed above (Schulz 2001).

Second, not only information technology can help in knowledge circulation, but it can also support knowledge production itself. In that case too, tools serve dedicated purposes. Kane and Alavi (2007) show that some tools (like knowledge-based repository and communication technologies such as e-mails) promote exploitation, while others (such as groupware and electronic communities of practice) support exploration. Tools supporting exploration are those that enable to maintain knowledge heterogeneity in the firm (a result congruent with March's original results). By contrast, tools supporting exploitation are those that contribute to the formation of a common knowledge base throughout the entire organization.

Accordingly, ambidexterity could be supported by a blend of different IT tools, some promoting exploration, others more exploitation focused. Fang et al. (2010) show that the optimal communication network regarding the balance of exploration and exploitation is made of which tightly connected subgroups with few ties between them. According to these authors, this configuration allows the firm to maintain a certain level of knowledge heterogeneity (for exploration) while guaranteeing a certain amount of common knowledge that eases coordination (exploitation).

3.2.2 IS for Exploration and Exploitation

IS for Exploitation

Exploitation activities produce knowledge that answers to concrete and immediate problems that arise in the organization's functions. This knowledge concerns best practices and is directly related to the organization's routines and processes (Turner and Makhija 2006). It is intended to be readily applicable in operations (Schulz 2001). Knowledge flows stemming from exploitation contribute to the development of a common knowledge base throughout the formal organization (Raisch 2008; Gilbert 2006; Taylor and Helfat 2009). In that sense, it contributes to integration at the organizational level and to the overall internal fit (Raisch et al. 2009).

Some information systems help in enhancing exploitation in organizations. Kane and Alavi (2007) show that knowledge repositories, because they provide the same knowledge to entire organization, contribute to the reduction of knowledge variance, the development of a common understanding and thus tend to standardize behaviors and responses to stimuli. Knowledge repositories thus contribute to a better exploitation of what is known by an organization by strengthening known routines and by providing a common language and understanding to every members of the organization. They further show that virtual team rooms, and more broadly project management tools, also enhance exploitation. In that case, knowledge variance is reduced within each project team. Although variance is maintained between teams, the lack of communication means prevent teams from using knowledge from outside their boundaries. As a result, each team builds its own common knowledge, isolated from the rest of the organization, and locally enhances exploitation.

From these results, one can deduce that all information systems that will contribute to the development of a common knowledge base, that will standardize and define fixed communication channels will contribute to a better exploitation of knowledge by organizations. Enterprise information systems, such as CRM or ERP, for instance, fall into that category.

IS for Exploration

Exploration, in essence, produces new knowledge from which organizations will be able to launch new activities remote from the competencies currently mastered by the organization. Explorative knowledge can thus in essence potentially change the structure and strategic positioning of the firm. Exploration entails the ability to fetch knowledge of great diversity, both within and outside the firm. Exploration also requires that the firm has the ability to confront, compare, and eventually combine these different pieces of knowledge in order to produce newness.

Some information systems may help in maintaining explorative abilities within firms. Kane and Alavi (2007) show that tools designed to create electronic communities of practice contribute to firm's exploration abilities. Because such tools are designed to allow people from everywhere in the organization to join a community of practice, they maintain links between different loci of knowledge, while always preserving the specificity of each community. Thus, electronic communities of practice combine both knowledge heterogeneity and connections, thereby increasing exploration.

Lindic et al. (2011) also show that web 2.0 tools, such as wiki, blogs, tools inspired from crowdsourcing, etc. also contribute to exploration. Because of their ease of use and implementation, these tools can be locally used for ideas generation, identification of weak signals provided by agents posting information from everywhere in the organization and identification of emerging trends thanks to the flexible format, allowing for the diffusion of all kinds of information. Moreover, these tools also share with electronic communities of practice functionalities that facilitate connections between agents. They thus help in developing new, informal networks around specific topics, thereby speeding up and giving more flexibility to innovation processes.

3.2.3 IS and Ambidexterity

Ambidextrous firms balance exploration and exploitation. Accordingly, one expects to find several types of information systems in an ambidextrous organization, some linked to exploitative activities, others supporting explorative activities (Kane and Alavi 2007; Magnusson 2004). Magnusson (2004) for instance, show that a suitable mix of explorative and exploitive IS could be electronic communities of practice coupled with more rigid knowledge directories.

However, Kane and Alavi (2007) show that there may be some interactions between the different kinds of information systems implemented in an organization. Moreover, not all interactions seem to be desirable: while some enable complementarities, others do not and are even detrimental to performance. In particular, testing different combinations in a simulation model, Kane and Alavi (2007) show that when a firm is rather oriented towards exploitation, then augmenting exploitative IS (such as knowledge repositories and virtual team rooms) by explorative ones (such as electronic communities of practice) is beneficial because it introduces flexibility and variety to an otherwise rigid system. On the contrary, in a exploration oriented organization, introducing IS that support exploitation may break the knowledge creation dynamics.

This suggests that, it does not suffice to implement both kinds of information systems in an organization. In addition, the appropriate mix of systems will depend on the firm's strategic orientation and core activities. This also suggest that in a complex organizations, made of several different parts, each facing specific tasks and sub environments, IS architecture should fit the organizational structure. In large organization having several lines of business, we find agents dedicated to exploration, others dedicated to exploitation, and specific IS should be used at specific times by specific agents facing specific tasks in the organization. This kind of approach has been taken by Lindic et al. (2011) in their study of the role of web 2.0 based tools in the innovation process. They propose that, at each step of the innovation process, specific tools should be used, depending on the task at hand. In order to identify the role of IS in the balancing of exploration and exploitation, it would thus be necessary to clearly identify the way firms actually achieve ambidexterity, and then to map information systems to the different components. This cartography, however, has yet to be established.

3.3 Managerial Practices

3.3.1 Complementing Organizational Design

Although there is a stream of research that has examined empirically the efficacy of the arguments made by knowledge based theories of the firm (e.g., Sabherwal and Becerra-Fernandez 2003; Lakshman and Parente 2008; Teigland and Wasko 2003), a central issue pertaining to organizational theory that remains to be convincingly resolved is that of appropriately designing organizations for maximising the benefits of knowledge management initiatives (e.g., Davis and Lawrence 1977; Galbraith 1973; Joyce 1986; King and Sethi 1999), including ambidexterity. Many organizational theorists have suggested, argued for, and empirically examined the usefulness of matrix organizational structures for effective handling of information and knowledge management in and around organizations (e.g., Burns and Wholey 1993; Katz and Allen 1985). Some researchers have suggested that executive leadership is as critical as organizational design to the effective management of knowledge in organizations (e.g., Viitala 2004).

A growing stream of research is pointing to the efficacy of knowledge management initiatives in organizations (e.g., Armstrong and Sambamurthy 1999; Lakshman and Parente 2008) and to the soundness of knowledge as a base of competitive advantage. The knowledge management literature has theoretically as well as empirically begun to identify both the internal and external (across organizational boundaries) processes of knowledge management (e.g., Alavi and

Leidner 2001; Teigland and Wasko 2003). However, this nascent yet burgeoning research stream tends to focus more on organizational processes of knowledge management (e.g., Lakshman and Parente 2008; Teigland and Wasko 2003) than on the structural issues surrounding effective knowledge transfer (see Lakshman 2008) in large and complex organizations. Although structural issues involved in coping with information-processing demands of organizations have been addressed in previous decades (e.g., Burns and Wholey 1993; Joyce 1986; King and Sethi 1999), their natural extension to knowledge creation, sharing, and leveraging has not followed, with very few exceptions (e.g., Reagans and McEvily 2003; Tortoriello 2008; Young et al. 2004). Even these exceptions focus on a broad range of organizational issues (not exclusively knowledge issues) (e.g., Young et al. 2004) and on organizational network theory (as opposed to structural) variables in the context of knowledge management (e.g., Reagans and McEvily 2003; Tortoriello 2008). Thus, even though past organizational literature would suggest examining matrix structures for maximizing knowledge sharing benefits (e.g., Burns and Wholey 1993; Joyce 1986), these structural issues are yet to be investigated comprehensively with exclusive attention to the knowledge domain. The case study presented here serves as a first step in this regard and serves as the basis for future structural investigations pertaining to knowledge management. We present the managerial practices of the multinational corporation (Legrand) including its use of a matrix structure to manage ambidexterity.

Researchers studying organizational design have suggested that matrix structures are team oriented arrangements that promote coordinated, multidisciplinary activity across functional areas, broad participations in decisions, and consequently the sharing of knowledge (e.g., Burns and Wholey 1993; Joyce 1986). Though the evidence is only partially supportive, this stream of research also argues that organizations employ matrix structures to increase the capacity for information handling, and enhancing participative decision making, by establishing formal lateral channels of communication, to complement and supplement existing hierarchical channels of communication (e.g., Davis and Lawrence 1977; Galbraith 1973). One reason why the evidence is only partially supportive is the resulting complexity, and consequent imbalance of power in the two arms of the matrix (e.g., Katz and Allen 1985), role overload, heightened role conflict, and political conflicts (e.g., Burns and Wholey 1993; Joyce 1986) which could potentially lead to the abandonment of the matrix. Some researchers suggest that a certain amount of conflict is designed into the matrix structure (e.g., Pfeffer 1981) and hence it requires productive management and careful institutionalization of a supporting culture that can potentially yield higher levels of coordination and decision making capabilities that result from the sharing of knowledge. Joyce's (1986) findings suggest the absence of evidence for role overload and role conflict, despite their dominant emphasis in the literature. However the evidence for the presence of political conflicts in abandonment of matrix structures provided by Burns and Wholey (1993) has lead to the general conclusion that the careful and productive management of matrix structures is essential for its success.

In this case study research, we focus on the broad set of managerial practices utilized by the organization in its efforts to manage the innovation process, including new knowledge generation and the exploitation of existing knowledge. These practices are embedded in fairly detailed processes that are integrated with the structural arrangements that facilitate these processes. We detail these processes in the section on findings.

3.3.2 Focus on Human Resources Configuration and Practices

The importance of human resource management practices for enhancing an organization's human capital and to equip it to handle organizational learning processes is fairly evident from the literature (e.g., Lepak and Snell 1999; Kang et al. 2007). It is also evident from the literature that these HRM practices are critical for building and maintaining the appropriate social and organizational capital (e.g., Ferris et al. 1998; Kang et al. 2007; Wright et al. 2001). Recent work has investigated the human resources architectures that facilitate ambidextrous learning (Kang and Snell 2009) and has identified configurations that facilitate exploratory processes of learning and those that facilitate exploitation. Based on the increasing availability of evidence that suggests that balancing exploration and exploitation is valuable for organizations and can contribute to performance enhancements, these researchers (Kang and Snell 2009) identify configurations of human resource practices that enhance the two processes of organizational learning, which are sometimes mutually incompatible and compete for the same organizational resources (e.g., Gibson and Birkinshaw 2004; He and Wong 2004). More specifically, this work has identified two alternative intellectual capital architectures that are well suited for ambidextrous organizational learning. The first of these two alternative architectures, labeled 'refined interpolation', depends on combining a specialist human capital, with cooperative social capital, and an organic organizational capital. The second architecture, labeled 'disciplined extrapolation' uses a combination of a generalist human capital, an entrepreneurial social capital, and a mechanistic organizational capital. These architectures and their suitability for ambidextrous learning are derived theoretically and are yet to be empirically examined. We provide an in-depth investigation of the human resource practices within an MNC and detail the interplay of these practices with other key functions involved in the development and maintenance of innovative knowledge stocks.

HR practices are contemporarily viewed as bundled configurations or sets (Huselid 1995; MacDuffie 1995) and are seen as being comprised of three distinctive types of practices, viz., development system, employee relations system, and performance/control system (Kang and Snell 2009). These three types of HR practices are conceived to have strong connections to the human capital, social capital, and organizational capital components respectively, of the overall intellectual capital of the organization. Within this system, human capital is seen as being managed by HR practices focusing on the identification of skill requirements, job specification activities, and training and job rotation, all of which constitute the development system. Social capital is seen as being managed by HR practices such as socialization, advancement and inducement, and attachment, constituting the employee relations system. Finally, organizational capital is seen as being managed by HR practices of job design, workflow, appraisal, and empowerment, constituting the performance/control system. Variations in each of these three systems of HR practices are seen to influence the three intellectual capital components in different ways and thus facilitate the processes of refined interpolation or disciplined extrapolation, with other configurations also being possible. However, practices consistent with either refined interpolation architecture or disciplined extrapolation architecture are best suited for ambidextrous learning. We examine the case of Legrand and study its HR practices to confirm their consistency with the above theoretically derived set of configurations. We describe our findings in this regard later in this document.

3.4 Differentiation Versus Integration

Recent literature seeks to reconcile different forms of ambidexterity. This problem has been framed in terms of differentiation versus integration dilemma (Raisch et al. 2009), after the seminal work of Lawrence and Lorsch (1967).

According to Lawrence and Lorsch (1967), differentiation corresponds to the state of segmentation of the organizational system into subsystems, each of which tends to develop particular attributes in relation to the requirements posed by its relevant external environment (Lawrence and Lorsch 1967). Each subsystem will be characterized in terms of structure, social concerns, time horizon, and goal, each of these dimensions being progressively informed by the specific sub-environment, itself characterized by its level of uncertainty, the subsystem faces. Brought into the realm of exploration/exploitation, Raisch et al. (2009) propose to call differentiation the subdivision of tasks into distinct units that tend to develop appropriate (internal) contexts for exploitation and exploration (Raisch et al. 2009). Clearly, from what precedes, differentiating exploration and exploitation requires adopting structural ambidexterity (Andriopoulos and Lewis 2009).

The interest of differentiation or structural ambidexterity in the exploration/exploitation literature stems from the fact that exploration and exploitation are seen as putting constraints on one another. Isolating these two activities in specific structures would thus permit to alleviate these constraints. Constraints can be imposed by exploitation on exploration. First, it might be difficult for a single unit to pursue both exploration and exploitation due to bounded rationality (Gilbert 2006). But constraints can also be imposed by exploration on exploitation. In particular, exploration shakes the organization and destabilizes exploitation activities. This is a serious concern because profits are generated by exploitation, and exploration can only be pursued if it can draw on resources earned by exploitation (March 1991). In addition, exploration distracts resources from exploitation.

However, isolating one activity from the other, conveys the risk for the firm to lose its internal fit (Siggelkow 2002). In some extreme case, the firm may end up with two fully duplicated value chains, each one operating on a specific sub-environment (e.g. electronic and paper-based activities for a newspaper; Gilbert 2006). So, when a firm resorts only to differentiation, each unit could stabilize in exploration or in exploitation, and the firm would somehow lose its overall integrity (Raisch et al. 2009).

It is thus necessary to combine differentiation with integration. For Lawrence and Lorsch (1967), integration is the process of achieving unity of effort among the various subsystems in the accomplishment of the organization's task. Scholars from organizational learning field adapted this definition to the specific context of exploration and exploitation and call integration, the behavioral mechanisms that enable organizations to address exploration and exploitation within the same unit (Raisch et al. 2009). From the previous discussion on the different forms of ambidexterity, it is clear that integration equates contextual ambidexterity (Andriopoulos and Lewis 2009).

It is important for a firm to have integration mechanisms because it is the mechanism through which the firm can deal with the entire environment, and not only a set of independent sub-environments (Lawrence and Lorsch 1967). For researchers from the ambidexterity field, integration is necessary because it is the mechanism through which the firm can combine new knowledge to the existing knowledge base. This relates to the notions of internal fit and coherence of the firm. It also conveys the idea of optimizing internal processes, and reducing transaction costs. So, while differentiation is underlaid by the idea that activities are decomposable, and thus that the constraints imposed by one type of activity on the other can be relieved, integration reintroduces the idea that these constraints exist and must be cope with. Raisch et al. (2009) go even further and assert that combining integration and differentiation is necessary for ambidexterity to really be a dynamic capability for the firm.

So far, works on differentiation and integration of exploration and exploitation have mostly focused on organizational configurations that allow a firm to host integrative and differentiating mechanisms. The bulk of the research seeks to determine whether differentiation should be strategic and integration tactic or if it should be the other way around (Raisch et al. 2009).

O'Reilly and Tushman (2008), for instance, suggest that top management should be the integrating lynchpin between structurally differentiated explorative and exploitative units. Jansen et al. (2009) found that integration between explorative and exploitative units are better made by formal mechanisms at the operational level and informal ones within the top management team. Andriopoulos and Lewis (2009) describe how integration and differentiation are nested mechanisms at the individual, project teams, and top management level.

Yet, if understanding the role of organizational design is necessary, it is, however, insufficient to fully account for how differentiation and integration operate in a firm. Indeed, a full understanding requires the integration of the time dimension in our theories (Schreyögg and Kliesch-Eberl 2007; Raisch et al. 2009; Schreyögg and Sydow 2010), leading to a focus on organizational processes.

Indeed, organizations continuously undergo emerging and unexpected events and must continuously adapt (Martin and Eisenhardt 2010). Moreover, exploration and exploitation activities themselves evolve and organizational change and adaptation are necessary as learning unfolds (Raisch et al. 2009). Therefore, it is likely that within organizations, the different actors adapt to these internal and external changes and modify explorative and exploitative activities, as well as the balance thereof. Ambidexterity can thus not be considered only as a matter of static organizational design but must also be seen as including a dynamic component. As Jansen et al. (2009) put it "*[...] organizational ambidexterity refers to the routines and processes by which organizations mobilize, coordinate, and integrate dispersed exploratory and exploitative efforts, and allocate, reallocate, combine, and recombine resources and assets across differentiated units*" (Jansen et al. 2009, p. 799).

Hence, we think it is important to introduce time in our theories on ambidexterity and to develop a process view on ambidexterity. A process view implies different phases in which differentiation and integration are operated and balanced differently. These different requirements call for different organizational contexts and managerial settings. From that standpoint, firms continuously and consciously tune their cognitive activities. These are local and conscious choices made by actors at different levels. The process also has to be consciously monitored both within each phase, and also to decide when to move from one stage to the other. However, given its dynamic nature and the fact that different actors and different settings may be required at each stage, this monitoring is likely to be local and handled by different types of actors at different organizational levels (Schreyögg and Kliesch-Eberl 2007; Schreyögg and Sydow 2010). As Raisch et al. (2009) put it, "*First, managing for ambidexterity is a task of dynamic rather than static alignment. Second, different solutions, including structural and contextual ones, may be required over time to sustain ambidexterity. Third, ambidexterity may arise from both simultaneous and sequential attention to exploration and exploitation*" (Raisch et al. 2009, p. 689). Thus, a process view enables us to see ambidexterity as sustained by related but heterogeneous arrangements, each sustaining different stages of the process and being situated in time, space and organizational loci (Schreyögg and Sydow 2010).

Yet, so far, this line of research is abstract and conceptual except several mathematical models suggested to analyze these issues basing mainly on the solutions obtained on an NK-landscape (Siggelkow and Levinthal 2003; Siggelkow and Rivkin 2005). It has been shown that the value of decentralization and the optimal amount of integration is contingent upon both, the amount of environmental turbulence and complexity faced by a firm. Nevertheless we need a more precise view of the processes at work (Schreyögg and Sydow 2010).

With that objective, we need at the very least to determine (i) how such a process might look like and what are the different stages of it; (ii) how each stage is implemented in terms of structure and managerial mechanisms; (iii) where in this process integration and differentiation takes place and what are the transition mechanisms between the two.

3.5 Summary

From our literature review, ambidexterity appears has having been mostly studied from an organizational design perspective. Though essential, this view is mostly static and focuses on the structures that enable ambidexterity. More particularly, we focused on two components of this structure, information systems and human resources policies.

However, we argue that we must introduce time and adopt a more dynamic view on ambidexterity, especially if one adopts Raisch et al. (2009) suggestion to consider ambidexterity as a dynamic capability. To do so, we believe that it is necessary to identify the process dimension of ambidexterity and describe the different stages thereof. Regarding information systems, it will then be possible to see what kind of information systems should be used within each of these stages, thereby mapping the different information systems to the ambidexterity process.

In the next part of this report, we will investigate this process view by resorting to a single case study in an MNC, the group Legrand. This company is a major player in the electrical appliances sector. This activity entails the creation of new products at a relatively fast pace, but also their mass production. We believe that the choice of this organization is suitable for our purpose because (i) as an MNC, it needs information systems for communication and coordination among geographically distant sites. The firm should thus be well populated with IS; and (ii) as a large organization from the industrial sector where both new product development and mass production are necessary, it should be concerned with the simultaneous management of exploration and exploitation.

4 Case Study: The Group Legrand

4.1 Context

The group Legrand is a multinational company created in 1860 and operating in electrical appliances industrial sector. In 2009, its annual turnover was € 3.6 billion and its net income was € 291 million. The company employs 30.000 people worldwide.

The group has manufacturing and/or distribution subsidiaries and offices in more than 70 countries, and sells its products in about 180 countries. Its key markets are France, Italy and the United States, which accounted for approximately 53 % of annual revenue in 2010 (2009, p. 56, 2008, p. 54 %), new economies keeping growing to account for one third of revenue.

Legrand has leadership positions including the number-one place for wiring devices and cable management worldwide, as well as number-one rankings on many key national markets. Its key markets are France, Italy and the US, and the company is at the front of the field in at least one of its business segments in nearly all Latin American countries, China, Russia and India. In Brazil, the company ranks first for wiring devices, audio and video door-entry systems and modular circuit breakers. The firm is leader in wiring devices and audio and video door-entry systems in China, and in wiring devices and cable management in Russia. Emerging markets make a growing contribution to the group's growth. Today they account for 30 % of sales—a proportion that has risen close to two percentage points a year since 2003.

Legrand has focused its acquisition strategy on local leaders in countries or market segments with high growth potential. Since 2005, 19 companies have been acquired, nearly half operating on emerging markets. Examples include HDL and Cemar in Brazil, Estap in Turkey, and TCL Wuxi and Shidean in China, most of them being frontrunners on their markets.

Three years ago, Legrand modified its overall organization. There is now a global organization of the back office with three industrial divisions in charge of the R&D and the production and local front offices in countries, responsible for sales.

The three industrial divisions are:

- ASR (Appareillage et Système Résidentiels): this division develops products for homes, such as switchgear and control gear and, more generally, any type of home systems. It represents 45 % of the group's total sales and is its largest division.
- Tertiary: equipments for offices in the tertiary sectors, telecommunication cables, electric systems, etc. It is the mirror of the tertiary sector and provides all electric appliances this type of activity may need.
- EDIA (Electric Distribution and Industrial Applications): this division develops power systems and associated control devices, such as circuit-breakers, cabinets…

The aim of this organization of the back office is to optimize for costs and investments. There was also a felt need to speed up the new product development process (projects cycle has been reduced from 4 to 2 years).

Front offices are usually associated with a specific country and are in charge of developing their specific market. Each front office is rewarded following the market shares gained, the increase in profitable sales and the optimization of the need in working capital. Legrand's clients are wholesalers. The group operates in the business-to-business sector and has no contact with its products end users. Although there seems to have been some experiments to open new distribution channels, business-to-business remains the dominant distribution mode.

In addition to the front and back office, there is a corporate structure providing different support functions such as purchasing, finance, strategy, design that are increasingly standardized throughout the entire organization.

Because Legrand is a multinational company that both relies on a high innovation rate and on frequent external acquisition, it is a particularly well suited case for our purpose. Frequent acquisitions of new companies raise the question of knowledge circulation in a potentially highly fragmented organization, replete with internal barriers to knowledge circulation. At the same time, however, because of the necessity to diffuse innovation, firms need to quickly spread knowledge stemming from exploration. One may thus expect that Legrand, given its success, has succeeded in managing integration and differentiation of exploration and exploitation.

4.2 Method

To gain insights into exploration and exploitation dynamics within an organization, we conducted a descriptive and explanatory single case study (Miles and Huberman 1994; Yin 2009). We have chosen a research strategy based on a rich case data as inspiration for new theory and ideas (Siggelkow 2007) concerning the coexistence of both types of ambidexterities within an organization.

In order to study how Legrand succeeds in managing the balance between exploration and exploitation in a geographically scattered organizational structure, we decided to conduct interviews in three different locations: the head-quarter in Limoges, France, the Russian subsidiary and the Indian subsidiary. We choose these two subsidiaries because of their different features. The Russian implantation is 13 years old. The Russian subsidiary does not have any research center and one production plant producing aging technology. The Indian subsidiary has been created in 1996. The Indian subsidiary hosts different engineering and design departments and has a premium market segment more developed than the Russian one. We thus expected to have contrasted situations and potentially different kinds of interactions between subsidiaries and headquarters.

Collection of the data presented in this report began in November 2010 and ended in March 2011. Data were gathered from multiple sources: (i) interviews of 10 managers; (ii) notes from direct observation (meetings and intern workshops); (iii) secondary data (internal and external documents: activity reports, press articles, etc.).

We interviewed people from the R&D department of the ASR (Appareillage et Systèmes Résidentiels) division, from the central marketing department, from the central HR department, former head of the Russian subsidiary, from the Russian HR department, from the Russian marketing department and people from internal communication department. This variety of points of view gave us a global overview of the firm's functioning.

Notes were taken during the different meetings and workshops with executives, and observations were written in a research diary, thereby providing real-time data (Miles and Huberman 1994). A first step of analysis consisted of comparing the real-time data collected to identify common dilemmas and to define the unique aspects of each of LEGRAND's organizational units. We created tables and graphs to facilitate further comparisons and compared successive structures for similarities and differences to refine the conceptual insights in order to develop the emerging constructs and theoretical logic. Then the committee identified key coordinators and managers to be interviewed. Interviews were recorded on tape and transcribed. They typically lasted 110 min, although a few ran as long as 3 h. Collected data were first validated by respondents and second by the knowledge managers and executives involved in the steering committee, as they are not necessarily those first investigated. Each author independently coded these data according to our framework, and coding differences were resolved through discussion. We added complementary information from the analysis of secondary data. As the analysis evolved, in vivo codes have emerged, inferential and explanatory (Miles and Huberman 1994) that led us to look back to our real-time data in order to be sure that our progress was coherent with them. This allowed us to stabilize our grid of codes, but also to increase significantly the correspondence between data blocks and codes, i.e. inter-coding reliability (Miles and Huberman 1994). The summary of findings was regularly submitted to the steering committee. Therefore, the use of multiple data sources, but above all our frequent interactions with the steering committee, significantly strengthened the quality of our findings.

4.3 Findings

4.3.1 Main Finding: Overall Innovation Process at Legrand

Our interviews reveal that innovation at Legrand can be described as a cycle involving the entire organization. In the upward phase of this cycle, Ideas and concepts for future products are generated throughout the organization, including operational levels in the different countries where the group operates. These ideas are then sent to headquarters, to the different divisions, where ideas are gathered, further assessed and developed. Projects deemed most promising are then submitted to top management that will decide on which projects to invest. Once this decision made, innovation enters the downward phase of its cycle. This phase begins with an exploration phase in which, R&D on new products is conducted centrally in some R&D centers, mostly located in France and Italy. Once prototyped, the products are then produced marketed and sold in the different subsidiaries, close to their markets.

We quote of the interviewee that summarizes this overall process: *"What you have to understand is that the development process has two major phases: before validation of investment by the CEO (we present every project with a cost*

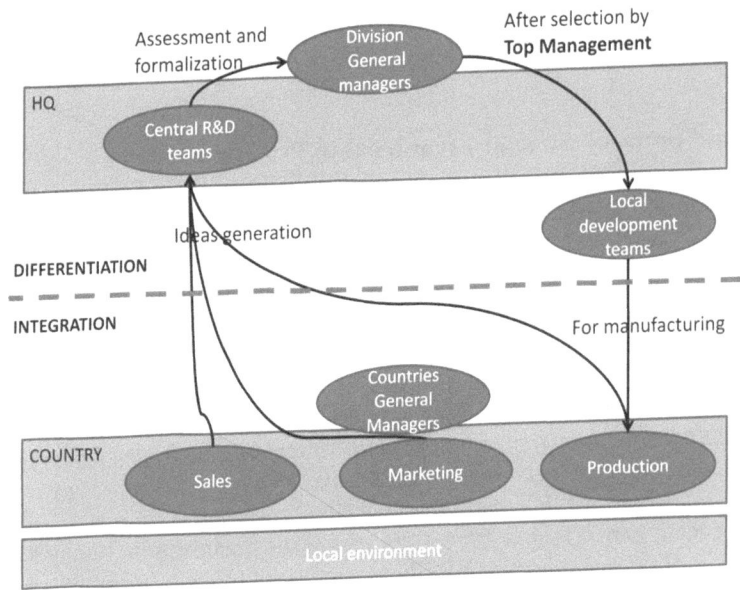

Fig. 1 A process view of exploration/exploitation management at Legrand

estimated over € 250.000 (which is almost always the case for us) to the CEO, we prepare the guidelines. For a range of equipment, we prepare its architecture, how we want to make it, how much it will cost [...]. This is made centrally. [...] Now we succeed our exam with the CEO. Now we are in the downstream phase. These are our Indian colleagues that manage the project" (Development, Equipment and Home System Division).

Our study reveals that different forms of ambidexterity are present at different stages of this innovation cycle. Contextual ambidexterity is visible early in the upward phase, when new ideas are generated. Structural ambidexterity is evidenced in the downward phase, where exploration is conducted in Europe in the main R&D centers, while exploitation is conducted in the different subsidiaries. Moreover, we find that ambidexterity at Legrand is made possible thanks to specific structural and managerial features. At the structural level, the firm set a number of common, standardized elements, such as a matrix structure, products platform, and standardized information systems. These elements ensure the smooth circulation of old and new knowledge throughout the firm. It is complemented by some managerial practices that ensure some flexibility in the system. In particular, internal mobility and the support of informal networks guarantee creative abrasion and cross-fertilization.

In the rest of this section, we will present our empirical materials following this global scheme (Fig. 1).

We will first present the upward stream, highlighting the different forms of ambidexterity mobilized and the structural and managerial settings that support them. We then present the downward stream, the ambidexterity involved and the means used to sustain it.

4.3.2 An "Upward" Stream: From Ideas to Projects Portfolio

Operational: Ideas Generating at the Operational Level

One fact stressed by many interviewees is that idea generation is encouraged throughout the organization. Whether in R&D, marketing or internal communication, whether at the headquarter or in subsidiaries, all the people we talked with stressed that they were encouraged to propose new ideas that could potentially lead to new products and/or new markets for the firm. Although not the only reason, this feat of the corporate culture undoubtedly contributes to its agility. As one of the interviewees asserts: "*I think we are agile, so we are going to seize things. I think 3 years ago, we would not have talked of energetic efficiency. Assisted living, 4 years ago, we would not have talked about*" (Internal communication department).

These ideas may originate from every part of the organization. It may be in R&D centers: "*In development teams, there are plenty of technicians, of engineers, people who may say at some point, "hey, what if we put leather [around switches]". We must then discuss if we have an idea, how to implement it. This idea, we are going to take it and there are organization groups, on Friday (where I come as a referee), they say we had the idea to put leather, is it interesting or not? If it is "no", because there is no business behind or we are going to put a patent on it to say we patented something and put it aside because there is no business behind at the moment. If it is "yes", we will put some money on it to dig the idea*" (Development, Equipment and Home System Division).

But idea may also come from subsidiaries: "*New product development stems when we are in a subsidiary, from local market needs that are not fulfilled yet. [...] There have been some attempts to propose entirely new products. New even in terms of conception. We do not master the decision to make or not these products. Hence, we gather information, we make a proposal at the group level. We say: "we think of such product. Would that be a good idea or not? We think it can bring an innovation or new market segment. We do not make that product today, but what do you think about it?"* (Equipment and Home system Marketing\

Besides, although this idea generation process stems from individual initiatives, they are encouraged and supported by middle and top management. Aware of the importance of such processes, the middle management provides spaces in which these ideas can be expressed and discussed. For instance, at the Equipment and Home system division, "*we have a meeting "staff and development", every Friday. It is the French division team, but all the persons in charge of the development in the group are invited. It is at 11/1\ am to have Chinese, Indian, Brazilian, Hungarian, Italian people. And then we have an exchange meeting. It is formal, we have an agenda, minutes,... but it's really about exchanging and sharing. [...] In each center, we have meetings where we are going to look what is new, what is* *being made. And, at some point, I will decide to discuss this at one of the Friday's meeting. It is very informative, sharing of best ideas...*" (Development, Equipment and Home System Division).

Moreover, there are several initiatives to use collaborative tools to communicate innovative ideas. "I will show you an example of SharePoint. I wanted it really simple. I called it *"it makes me think of..."*. *I thought that each person working in an enterprise has a brain that works by analogies. We are on a weekend, for instance, we are going to watch TV, walk in the street or on the beach... and suddenly, you see something that makes you think of your professional life. I thus thought to open a space where to very easily tell what has made us think of something*" (Internal Communication Department).

For idea generation, the firm truly resorts to contextual ambidexterity (Gibson and Birkinshaw 2004). First, the trade-off between exploration and exploitation is resolved at the individual level. It appears that new ideas are generated by individuals, during their work time, in the course of their daily activities. Individuals, in this respect, thus decide how to manage their time and allocate their efforts to exploration and exploitation. Moreover, the organization creates and maintains a favorable context within which individuals can be ambidextrous. Besides the production objectives, individuals are encouraged to produce new ideas, and this effort is valued by the organization, through collective meetings, patenting, etc.

Considering the individual level has two consequences. First, because of cognitive constraints, most of the individuals are only aware of their local surroundings, and their exploration or exploitation efforts mainly aim at shaping their immediate environment. Second, most of the agents do not have the power to act directly upon the activity system because they lack ownership or user rights to significantly modify activity chains (Stieglitz and Heine 2007). As a consequence, individual exploration/exploitation mainly deals with knowledge and ideas. Individual agents seek to develop local knowledge base and make sense of their immediate surroundings. Hence, contextual ambidexterity is used for idea generation. Up to this point in the innovation cycle, no significant investment has been made by the firm. Only in a second step, when new knowledge will be stabilized, will they enter in a communication process to convince decision-makers to change the activity system following their recommendations.

Middle Management

It is widely acknowledged that middle management plays an important role in knowledge creation and knowledge transfer (Nonaka and Takeuchi 1995). Middle managers occupy a unique position between top management that set the general strategic orientations of the firm and operational levels at which, daily, tactical decisions have to be made. Middle managers are thus ideally positioned to translate ideas and knowledge from one world to the other. Middle managers are able to see both the strategic interest of an initiative, while being aware of implementation requirements. They are thus able to have a complete view of the process and stakes of a new project (Martin and Eisenhardt 2010). In this study, we call middle management general managers that are running countries or managers who are responsible of departments in divisions and are located at the headquarters, in France.

In the group Legrand, middle managers see themselves as enablers of innovation and consider that one of their important roles is to spark innovative ideas. As one of the interviewees explains: *"One cannot say that corporate culture encourages innovation. Now, the group functioning is one in which people have great autonomy. With this autonomy, when you are for instance head of a country such as Russia, you have annual objectives (turnover, yields ...) and you are required to meet these objectives, without being told how. In your job, you have a high degree of improvisation, but that won't be acknowledged. It de facto exists"* (Russian subsidiary).

One can identify two kinds of middle managers, occupying different positions in the organizations. Some are heading departments within divisions (such as marketing or development), while others are general managers in charge of countries. Middle management in central departments within divisions, at headquarters, consciously seeks to promote breakthrough innovations. *"[...] we have what we call radical innovation. It's quiet new. We started about 1 year ago to say that even if spontaneous innovation is good, we need to think of tomorrow's concepts. And I am in charge of that, delegated by my boss to do that for the entire division. That is to say, I am in charge to find systems (that's how I call them). [...] There we created a think group (these are not people working full time on that) in which we say we must think of tomorrow's home electric system (home, it's my job). [...] It will bring advantages, namely mobility (move the bed for instance and realize that there's no plugs. And that, tomorrow, our ambition is to be the first to propose answers to that. This is radical, these are objectives that are: give end users solutions that we will be the only ones to bring in terms of flexibility, moving beds, TV ... There I wouldn't have to make a hole or to call an electrician to fix three plugs (it's complicated, and electricity we don't like much). This is radical [...]"* (Development, Equipment and Home System Division).

Middle management in subsidiaries too seek to promote radical innovation and breakthrough ideas. For instance, *"In Russia, we work a lot in the industrial domain [...]. In Russia, on working site, we have to resist temperatures ranging between −40 and +40 °C. European products, it's −10 °C maximum. To go to Russia, we are going to say: make us a product that can resist −40 °C. Yes, countries ask either adaptations [of existing products] or new products"* (Russian subsidiary).

Thus, middle managers occupying key positions in the organization seek to leverage the spontaneous process of ideas generation occurring at the operational level. First, they rationalize and organize it by setting specific work groups dedicated to the generation of new ideas. In so doing, they increase the power of the organization for new, far-fetching ideas. Second, they formalize ideas stemming from operational levels and send them to higher hierarchical levels in the form of new products needs. This is congruent with the role assigned to middle management by Gibson and Birkinshaw (2004) who see middle managers as important driving belt between strategic and operational levels. Our observations also echo

Martin and Eisenhardt (2010) who see middle managers increasing the resonance of micro phenomena occurring at operational levels. Middle managers are thus giving impetus to small, local initiatives and give them an organizational dimension.

However, having ideas generated, leveraged and formalized is not the end of the process. In order to become projects, these initiatives need to be funded. Investment decisions, beyond a certain level can only be made by the top management.

Top Management: Decision on Projects Portfolio's Structure

One of our interlocutors presents the top management team in the following way: *"we have people with background in engineering, in finance, in marketing. That's how is made the top management team. Younger people are 45–48 years old, older people are 63–64 years old. Women are rare. If we look at the executive committee, there are French, an Italian, an American. If we enlarge to the Monthly Industrial Committee, there are some more Italians and a woman, that's me"* (Internal communication department).

Top management has an important role regarding innovation and the development of new areas for business. These decisions are about external acquisitions, but also about organic growth, through internal innovation. Regarding external growth, *"we are very pragmatic, plenty of things come down from the CEO, I would say. I think of the strategy, of divisions and heads of subsidiaries [countries] part of whose mission is to spot potential acquisitions. It is crossed. In the end, decisions will be made by the CEO or vice-CEO but they are fertilized by lower levels"* (Internal communication department).

Top management also plays an important role regarding internal innovation. *"Top management is very important. In the growth, in entering new markets, impulsing new subjects. We are a very hierarchical organization in the sense that we are very disciplined. When they say so up there, we do so. At the same time, we are rather simple. That is, if the CEO calls one person, he won't go along the hierarchical line. Conversely, we can call him. His line is not filtered. Anyone can call him. If he's there, you get him"* (Internal communication department).

As mentioned earlier, every project beyond 250.000€ has to be agreed by the CEO prior to be launched. Innovation is one of the core values of the group and its importance is regularly asserted by the CEO. Top management is thus very involved in the innovation process. Each project is submitted to the top management and the management has to decide whether to launch a project that can be an innovation or an acquisition. The top management is in charge of managing the global technological and markets portfolio of the firm. Its decisions are about the structure of that portfolio and to balance the ratio exploration exploitation at the firm level.

Summary of the "Upward" Stream

What we termed upward stream is the process by which ideas are progressively hardened to eventually become technological projects. Ideas may come from any part of the organization. The middle management gathers, makes a first assessment and leverages the ideas deemed most promising. Once formalized, these proposals are then submitted to the top management that decides which project will be funded.

Exploration and exploitation are balanced at every stage, though in a different manner. During the idea generation phase, exploration and exploitation are balanced in a contextual way (Gibson and Birkinshaw 2004). Innovative ideas emerge spontaneously in the course of individuals' daily activities. It is thus at the individual level that decisions to spend more or less time and efforts to the development of these ideas is made. This is done in a context favorable to this kind of behavior that is encouraged by middle management, but also top management through the claim that innovation is the core of Legrand's success.

In the phase of idea gathering and strengthening, exploration and exploitation balance becomes more collective and formalized. It begins in work groups and internal seminars and implies the active involvement of the middle management. The process is however not completely formal. Work group are rather *ad hoc* and is not part of an institutionalized process. It is rather of the middle management initiative, each manager handling this stage in his/her own way. As time unfolds, the process becomes a project and thus increasingly controlled and formalized.

To the contrary, the investment decision made by the top management is very formal and is part of a highly institutionalized process of projects' review. At this point, the balance between exploration and exploitation is rigorous and standardized.

Once the portfolio set, projects are implemented and developed. This is the beginning of the second phase of the overall cycle we termed "downward".

Use of Information Systems in the Upward Stream

We have been able to identify the use of information systems to some—but not all—phases of the upward stream. The use of specific information systems is mostly visible in the phase of ideas generation. The functional information systems used by individuals in the course of their daily activities are supplemented by more collaborative tools, such as SharePoint, to sustain exploration. The firm dedicates a lot of efforts to the development and maintenance of interrelationships within the firm. For instance, *"in the export department, we gather heads of subsidiaries, sales directors and marketing directors so that best practices of one country are known to others. It opens up the mind of certain collaborators who work in a specific environment, with a specific positioning, and suddenly, they realize that they have not exploit a direction and that another country exploit it in a certain way and that maybe could also be done in their country. [This kind of meeting] is encouraged"* (Russian Subsidiary). The development of this dense network of relations in turn sustains cross-fertilization and idea circulation. Formal and informal networks are truly part of the culture of the enterprise.

In terms of ambidexterity, this means that the firm supplements exploitative information systems with more explorative ones. At their individual level, agents can then opt for the system the most suitable for their current task, either exploration or exploitation. For this specific phase of idea generation, then, the firm should leave at agents' disposal both kinds of systems.

When ambidexterity become structural, that is, when exploration and exploitation activities have been turned into specific projects, there does not seem to exist dedicated information systems to specific activities. That is to say, tools used by both kinds of project teams seem quite similar. We may however suspect that teams do not use the tools in the same way and that they offer enough flexibility to be used in agreement with the task at hand. Taking SharePoint for instance, the software most used in project management at Legrand, it can be used as a platform for developing communities of practice, but also as a virtual project room. Thus, given the flexibility offered by this kind of web 2.0 tools, one can expect them to be suitable to serve different goals and to support different governance modes.

On the contrary, the firm does not have any system to help in gathering, screening, selecting ideas, neither at the middle management level nor at the top management one. This means that there are no automated means to submit ideas to the appropriate level of decision, and that ideas are channeled rather informally, through interpersonal networks or via traditional communication means such as e-mails. Moreover, this also means that there are no repositories of ideas, nor any systems that could help in enlarging the attention capacities of managers. One might thus suspect that some ideas maybe left unexploited and that some opportunities may be missed by the firm.

4.3.3 The "Downward" Stream: From Labs to Markets

Exploration Conducted Centrally

Exploration is mostly conducted centrally, within the centre of the different divisions. For instance, in the case of the ASR division (Appareillage et Système Résidentiels), producing switchgear, controlgear and home systems, "*There are two leading teams in the division: a French team for the Legrand brand, and an Italian team for the BTicino brand [these are the two historical and star brands of the group]. These guys are in charge of the standing of the brand. [...] In France, we also have a laboratory. The laboratory is multidivision and I run it because I was its biggest client*" (Development of the ASR division).

This centralization of the R&D teams is also true in the other divisions. The main benefits that are derived from this centralization and pooling of resources seem to be the ease of coordination, as all actors involved are located in the same place. The interviewed person explains: "*For upstream validation phases, prototyping, ... I am completely autonomous. I can be hyper reactive, I can go very fast because I know how to make the design and send the model. The next day it is on the machine, the day after I have the products. I have all the trade associations in my perimeters, design, manufacturing, modeling, tools realization, laboratory, simulations,... I can do everything. We have a beautiful tool that allows us to be efficient [...] I know when I design how I will manufacture. This is really a strong advantage for Legrand*" (Development of the ASR division).

Once this phase of prototyping is over, the project is then passed on to downstream development teams. There are teams in France and Hungry working on the European Union perimeter, plus one in Italy dedicated to the BTicino brand. There is one team in Brazil, one in India and two in China: one focused on the Chinese market and the other for export toward China.

"Hence, preparation of the file up to the validation by the top management. This is rather steered by central teams in France or Italy […] We do the prototyping, core functionalities of products, we set the industrial scenario, we deduce the production costs, investments and a ROI. […] [The CEO] says OK it's a good project, you can go on. Project's responsibility is then transferred to local teams and we move on projects. And, every week, we will have meeting between central and local teams to share problems" (Development of the ASR division).

Once the new product is entirely set, with all its forecasted functionalities implemented, the firm can enter the production phase. This phase is highly contingent and may depend on costs, customs taxes, convenience considerations. Depending on these different variables, the production may take place in the country in which products are intended to be sold or abroad. The important point, however, is that this phase also give place to important knowledge exchanges and collaborations. For instance, "*[In a Russian plant], we build molds [for making new switches]. There, there are necessarily some exchanges. For instance, people from the Italian R&D teams come to [the plant] to see how we could build the molds. In the case of industrial exchanges, the EU comes and see what can be done or not in the [Russian plant]*" (HR, Russian subsidiary).

Since the R&D process is organized that way, the new product development process, from the origination of the idea to the moment when wholesalers have products in their stocks has been reduced from 4 to 2 years.

To Succeed in Transforming from Exploration to Exploitation

The fact that the firm is successful while centralizing its R&D is somehow counterintuitive. As noted by Martin and Eisenhardt (2010), for instance, conducting products innovation centrally entails the risk to come up with product that would not fit local needs. The probability of mismatch between innovation and market needs is increased. Further, central innovation requires transferring significant knowledge to production units and sales teams. There should then be huge coordination and communication costs associated with central innovation. Yet, Legrand had outstanding financial results in 2010. This tends to show that the firm has set specific and efficient structures and processes to route new knowledge from labs to production plants and to transform explorative knowledge into exploitative knowledge. From our interviews, the firm relies on a specific organizational design and on specific managerial practices to succeed in this transformation.

Organizational design

– A matrix structure

Legrand adopted 3 years ago a matrix structure: *"We have in parallel divisions and subsidiaries (better characterized as "countries"). Within subsidiaries, one finds industrial and sales activities. The industrial part is truly matrix-like and depends of the division, while the commercial part is managed by countries only"* (Internal communication department). Within the division, one finds all the functions of the organization (HR, marketing, development, manufacturing, etc.).

Overall, the matrix structure, and especially the divisions contribute to thicken the communication network of the organization. This organization facilitates communication and knowledge exchanges between the different functions involved in the process in bringing new products to local markets. It also helps in deploying knowledge quickly in different subenvironments. For instance, *"In [a particular technological domain], standards are similar in switchgears and control gears from one country to another. Hence, the product can be made for several countries. [...] The fact that divisions are international will help the R&D's boss and the division's boss impulse things elsewhere and make others benefit of all that. Our engineering departments are platform departments that work on subjects rather than countries"* (Internal communication department).

As a result, there are strong interactions between teams located in headquarters and those located in the different countries. *"At every level of the enterprise, you'll have relationships [between subsidiaries and] head quarter. When you are in a country, at the marketing level, to propose a new selling approach, a new orientation,... you have teams in relation with teams at the head quarter"* (Russian Subsidiary). Such exchanges also take place in support functions, such as HR or external communications.

- Process rationalization

In addition to the adoption of a matrix structure, the firm also engaged in a rationalization of their processes by introducing modularity in products development. Modularity presents the main advantage of enabling the re-use of stabilized components in the development of different products. In addition, because interfaces between the components are standardized, it allows modifying some parts of a system while leaving the rest unchanged. As one of our respondents explains, *"Regarding flexibility of our production, the idea of platforms that we implemented 3 years ago is an example. [...] The idea of platforms is to have in Europe ten ranges of products, for instance, with the same components. [...] To sum up, the system of platforms allows a flexibility, gains in productivity and gains in globalization. [...] Platform, it means that where I had five European processes yesterday, I will have only one tomorrow. Inside, I have common components to optimize for production costs and future development costs*" (Development of the ASR division).

As a result, it facilitates the transformation of explorative and exploitative knowledge by limiting the amount and complexity of knowledge that has to be transferred (Sanchez and Mahoney 1997). It also lowers the frequency at which transformation has to occur (Tran et al. 2010) and thus limits the risk of over-whelming production units with new knowledge.

Managerial practices

- HR

Our findings reveal that the approach of Legrand is clearly one of disciplined extrapolation. This HR architecture of disciplined extrapolation is one of two identified by Kang and Snell (2009) and involves combining a generalist human capital, an entrepreneurial social capital, and a mechanistic organizational capital. This HR architecture, in conjunction with the matrix organizational design, aids both the upward and downward streams of the innovation process detailed by us. The human capital at Legrand is primarily generalist with a predominant use of expatriates in the MNC's subsidiaries, combined with other host country nationals for support. The social capital at Legrand is primarily entrepreneurial, with his-torically significant levels of autonomy to the different country subsidiaries, which is now being balanced with higher levels of standardization. Still, there remains a sufficient level of autonomy in the hands of the country heads to contribute to balancing exploration and exploitation at the global level. The organizational capital at Legrand is primarily mechanistic, with clearly defined procedures, suf-ficiently centralized, and systematically pursued by the corporate office in ensuring that innovation yields significant financial results.

The Legrand group engages in a number of human resource management practices to work in conjunction with broader initiatives for managing the ambi-dexterity of the global organization. The entire gamut of practices range from integrating acquired units in various countries into Legrand's operations, domestic and international mobility of executives for knowledge sharing purposes, talent management using the *Talentis* e-platform, training and development for managers and lower level employees, performance management practices of key executives across country borders, and the continuous and ongoing efforts to standardize the HR practices across subsidiaries. The group strives for consistency across these HR practices to enable overall organizational objective accomplishment. Post-merger integration efforts are particularly challenging, given the number of acquisitions made by the group in multiple countries. In the words of one interviewee, "*It takes a long time to integrate an enterprise into existing operations. In the area of Finance, for example, there is no longer a difference between the corresponding units of Legrand and Kontactor (a Russian subsidiary)*". Culturally, there is an enormous gap between the approach of the group and that of Contactor. It is the equivalent of a 100 years of difference, they don't work in the same manner.....yes, certainly because there are three broad dimensions: administrative (HR/finance), commercial, and industrial (with engineers). It is also geographic. I discuss with people in marketing because they are in the hall just beside us whereas I don't even know where BE is located and so I wouldn't discuss with them. It's silly but informal exchanges do count. (HR, Russian subsidiary).

The organization envisions the HR unit as being the true link between the HQ and the different subsidiaries and there is an increasing effort on the part of the International HR group at HQ to standardize HR activities across subsidiaries in different countries. Such standardization is aimed to be in balance with the necessary levels of autonomy and country specific needs and therefore is progressing at a slow but steady pace. Such a balance between the standardization and local adjustment in HR practices gives the global organization the required mix of centralized and decentralized set of practices to create a truly transnational organization. One interviewee, a former head of the HR unit in Russia, who now works in the International HR area at HQ, stressed the importance of this increase in standardization to balance the earlier existence of higher levels of subsidiary autonomy, as the company is gearing up to better match the environmental demands of balancing radical and incremental innovation in products and services.

The overall HR policies are exemplified by the following quote: "*For example, the HR function has organized itself in the last 6 months to become a true link among subsidiaries. For the last 11 years I have been away from the headquarters. So, today we are demanding more and more of the subsidiaries that they do their personnel reviews on Talentis, to complete a report every trimester, to really work with HQ in exchanging with them. Effectively, we are standardizing the personnel review process, the reporting, certainly the internal mobility procedures.....In HR, we are going to stress on it, in finance it's already been done, in logistics, we are getting there as well. And that reduces the discretion of the country head? Yes. If all their direct reports (N-1 level) were subject to these procedures in their respective domains, he would be obligated to respect it. Clearly, in the HR unit, I had very little contact with the HQ, plenty of liberty in the matter of training, remuneration. My boss had a significant level of liberty. Now, even in the area of internal mobility of key people in the subsidiaries, we are required to refer to higher ups in the group*" (HR, Russian subsidiary).

The group is also trying to optimize the conduct of R&D and the development of technology in few centralized locations around the world, and then to exploit them in the rest of the countries. In addition, there is a constant effort to share and internally benchmark best practices throughout the world in different functional departments. Most of the group's work is done in project teams. Speaking on these issues and providing an example of the transfer of knowledge on best practices, one interviewee said, "*It's in process. That had been controlled when I left in June. There are Legrand products made at Kontactor but there's no transfer of technology, but it's assembly. They are components made in the EU and sent to Russia to be assembled.....In all the departments whether they are in Russia or here, we work in project mode. We have many meetings which bring together people from different functions and sectors around a common theme. The exchange on best practices is therefore facilitated. I, for instance, I lead an international work group on recruitment processes which respect the principle of non-discrimination. There is an American in the USA, an Italian in Italy, a Turkish person in Turkey, a French person in France. We do it all through videoconference or conference calls. During the discussion, we have multiple sessions dedicated to best practices on the subject. It's a way to circulate best practices*" (HR, Russian subsidiary).

-Internal mobility (long term, short term)

At Legrand, performance and talent management using the Talentis platform, eventually leading to internal mobility of executives plays a key part in managing the knowledge flows. In the context of succession plans and individual preferences of key executives being considered for internal mobility, decisions are made on optimal choices among different propositions. The group resorts to external recruiting only when all internal possibilities have been exhausted. This ensures that key executives with crucial knowledge of content and processes are utilized in locations where they are needed the most. Thus, this serves as a key component of knowledge flow and transfer initiatives, in line with broad organizational objectives identified in annual strategic plans. This is highlighted by one top HR manager at HQ: "*The subsidiaries come at the end of each year to present a budget to top management at HQ. There's an envelope (budgetary) of all employees, and they go on to manage their thing. For central functions, strategy, the HR people, top or central positions, we have the mobility platform mobilized every month by the HR people (it's for internal mobility). There they are placed in common meetings lasting 2–3 h (depending on situational factors) and top management is deeply involved in it. They consider all positions to be filled, the succession plans…all that is anticipated, and then the internal mobility is effected, we have individual interviews each year end where we see if people are inclined to move or not; if we want to have them moved because we think that it is best, but they may think otherwise. People are therefore passed through the review process and we try to examine possibilities. When we don't have people internally, we look outside. This is done transnationally, the finance mobility platform considers the entire population of finance people. It's done by discipline. Sometimes it is regrouped a little, as for instance, commercial and marketing*" (Internal communication department).

Consistent with their approach to optimize knowledge production and exploiting existing knowledge as much as possible, the group aims at managing their executive human capital to obtain these broad objectives. This translates into the utilization of expatriates from HQ in a number of subsidiaries, however, after having considered the possibility of retaining the host country executives heading the acquired entities. As has been identified earlier, the group has grown primarily through acquisitions abroad. Even when these expatriates are used, they are typically complemented by host country executives beginning from the second (from the top) hierarchical level. This blend of executives from HQ and the host country, and in some cases, regionally sourced top managers (as country heads) provides the blending of global and local knowledge, and familiarity with the internal organizational hierarchy at HQ, to effectively carry out the task of growth through innovation. Again, in the words of one top HR manager, "*We have a certain number of subsidiaries headed by French expatriates, for many cases, this is true. We also have local top managers. When we make acquisitions, we leave the top managers in place, except when they want to leave. Because, on occasion, host country nationals at the top want to leave. And we have people, I think, for example in our group of directors in France who are on the board of directors (he is French, but we acquired this company in France). The four Turkish subsidiaries are headed by Turks, the country head there is Italian*" (Internal communication department).

At Legrand there is a judicious blend of staffing, talent management, and internal mobility practices, as exemplified by the above interview responses. In addition to the movements described above, there is also a certain extent of movement of executives between subsidiaries. This is enriched by many executives travelling on short term missions to subsidiaries in other countries, with about an equal number of inpatriates on short term missions at HQ. All of these practices in general help knowledge flows between the headquarters in France and the multiple subsidiaries, and eventually to countries where there are no subsidiaries but commercial operations do exist. The following quote from a former top HR manager from Russia speaks to this issue: " *Internal mobility? It exists. It is fairly well developed and actively promoted. It works very well in France, there is mobility within work disciplines. Internal mobility on an international scale exists as well, but as it is expensive, it is less frequent. We have inpatriates who come to work at HQ (exclusively in Paris) and we have transfers between subsidiaries. But often, we stay within the same continent. It's simple that way, …..but there are a lot of people on specific missions in Russia for specific projects (transfers in production at Kontactor). We have people from France and from Italy every month*" (HR, Russian subsidiary).

-Training

In the area of training and development, Legrand uses a mixture of programs focused on broadening managerial and employee competencies on the one hand, and those focusing on enhancing functional skills, on the other. Such an approach is recognized to be the best in the management of ambidexterity objectives, requiring both exploration and exploitation of knowledge. Such an approach is also effective within the mixture of structural and contextual approaches to ambidexterity that we have identified to be in use at Legrand. Speaking on this issue, a senior manager in charge of International HR at HQ had this to say, "*Both. We have a lot of training programmes in Russia. We have training programmes on sales techniques, time and priority management, how to make presentations, etc.…...we also have group oriented training programmes by discipline, training on Legrand products, training on Legrand sales arguments, marketing programmes made for Legrand (practices and products of the group)*" (HR, Headquarter).

-Key people

Consistent with strategic human resource management approaches to managing key people, Legrand's practices in this regard ensure the optimal balance between new knowledge generation and exploitation of existing knowledge. Within the overall configuration of HR practices described so far, Legrand manages the recruitment of its key people across the world up to a certain hierarchical level in the organization. A senior HR manager says, *"The HQ is involved up to a certain level (N-1, sometimes N-2). They have their view, for recruitments for instance, they use the internal mobility platform in the course of an established process"* (Internal communication department)

-People review

The performance appraisal and management system at Legrand is based on a mix of both behavioral and results evaluation. This is once again consistent with the HR architectures best suited for ambidextrous organizations. The annual appraisal interview serves both as a stage for setting subsequent year objectives, as well as evaluating objective accomplishment for the most recently ended year. The following quote provides details of such a process of people review. "It is a mix of both. In the annual appraisal, there are specific objectives, with specified weights (totaling 100 %); then during the appraisal interview, there is a quantitative evaluation (to what extent objectives have been attained) and narrative commentaries on performance. Even with 0 % of the objectives accomplished, the evaluation could be positive." (Internal communication department).

-Sustain/encourage informal networks

In addition to all of the above mentioned specific HR practices, the group also places emphasis on building and sustaining informal networks of the key people around the world, both face-to-face and technological through the intranet. These serve as effective mechanisms for circulating best practices. The following quote from a senior International HR manager provides an example of these initiatives. *"Every year, every 2 years, every 3 years, there are seminars, gatherings, meetings organized for all functions. By functional area, we take all department managers from all subsidiaries. That lasts 3–4 days. There are transfers from HQ to subsidiaries but it happens that we ask certain subsidiaries to present what they do or do not do. These are moments where there are informal exchanges (it could well be an exchange on practices)….. The support for internal communication are brochures. As everyone reads them, there is a sharing of information. The intranet is important as well. That gives ideas"* (International HR).

To Realize the Fit with Local Markets

It is a truism to say that multinational corporations must combine global coordination and local responsiveness to specific subenvironments (Bartlett and Ghoshal 1988; Gupta and Govindarajan 1991; Gupta and Govindarajan 2000; Zellmer-Bruhn and Gibson 2006). This idea closely echoes the notion of differentiation versus integration proposed by Lawrence and Lorsch (1967). So far, we have dealt with the integration and global coordination aspect, presenting how firms manage to develop innovation and route them through to their different markets around the world. However, each subsidiary faces specific context and has to adapt to local contingencies. Local responsiveness implies specific marketing and distribution systems to cope with local laws and customers preferences.

At Legrand, this local responsiveness is allowed by the great autonomy lend to countries. Countries are thus allowed to develop the specific systems they need to better serve their customers. For instance, in Russia, about a new Do It Yourslef line of products, "*with import taxes, we could not import products from France. To bring them to Russia is much more expensive, and we would have a responsiveness problem. A shop never orders the same thing from 1 week to the other. They have stronger requirements in terms of delivery time than professionals and there was just a high risk not to have the good products in stock. Moreover, packages have to be written in Russian for end users. We would never have had an adapted packaging if it had been done centrally. Hence, it was easier to do it locally. We saw orders arriving; we took products from stocks [for B to B]; we packaged them. We were sure to have the correct indications in Russian, everything clean. And it was in due time in shops. There was a really high responsiveness, with a simpler packaging, adapted to Russian shelves. It's cost reduction on the packaging, delivery time reduction because we are faster in serving our customers. And better margins because we can set a correct price when we control costs. Thus, it is very interesting*" (Marketing, Russian subsidiary).

Not only there are important local adaptations to packaging and marketing, but also in production processes. "*We did a lot on production lines because producing locally is an important advantage on the Russian market because of transportation time, custom taxes, and taxes in general. Enhancing production lines was a permanent challenge in order to continuously enhance costs and being able to be as close as possible of the market price while keeping an interesting profitability*" (Marketing, Russian subsidiary).

How does the firm manage to maintain the autonomy necessary to local responsiveness? The first elements stem from the growth model of the firm. Because subsidiaries are usually created by external acquisition, the local entities have their unique culture and production and marketing systems. Moreover, the main driver of the decision to acquire a new subsidiary is to access market shares. This means that cultural or technological proximity do not enter into considerations for the decision to make an acquisition. As a result, the newly acquire firm is completely different from the rest of the group and remains *de facto* autonomous. "*Integrating an enterprise is a very long process. [In Russia, acquired in 2007], there are no more differences in finance [between the subsidiary and the rest of the group], nor on HR, but, typically, there is a huge gap in R&D. Culturally, there is a huge gap between the group approach and the one of the engineering department [of the Russian subsidiary]. They are 100 years late, they do not work the same way at all*" (HR, Russian subsidiary).

However, autonomy is not explained by historical reasons alone. There is a conscious willingness of the group to preserve autonomy of subsidiaries. This is most visible in the latitude left to heads of subsidiaries in the way they run their business. *"Each subsidiary (except for financial reporting for which there are precise instructions) is relatively autonomous. Rules followed in each subsidiary is set by its director. He is like the owner of an SME. He is required to reach a given result at the end of the year. After that, he has a lot of freedom about the way he reaches the required results"* (HR, Russian subsidiary).

This autonomy of subsidiaries' heads is exemplified by the following quote: *"in western Europe, saving energy is fashionable. Hence, there is a certain number of products that have been developed and some teams wished us to sale these products in Russia. I considered that considering the investments required, the size of the segment at the moment in Russia and my current priorities, I would not do it. Clearly, I had people here who made presentations saying that Russia has a 10 million euros potential within 3 years in this area and that the head of the Russian subsidiary would not put a cent to develop this activity because I consider it is not opportune"* (Russian subsidiary).

Hence, it clearly appears from this quote that heads of subsidiaries play a key role in the diffusion of new products. Moreover, within their countries, directors are free to organize their activities and the allocation of their resources: *"To take the decision to commercialize a new product in a territory that lies in a subsidiary's director is 100 % of his/her responsibilities, s/he does what s/he wants. [...] If, for instance, with the different directors of the CIS (Community of Independent States), I was selling a product in Russia and wanted to sell it in Uzbekistan or Kazakhstan, I did what I wanted"* (Russian subsidiary).

In addition to this great autonomy left to heads of subsidiaries, there are also strong internal knowledge flows and communications within subsidiaries. This is visible in the mobility policy set in subsidiaries. *"In Russia, in terms of mobility, we did it among juridical entities. For instance, we had a person in charge of logistics [in a plant] in Moscow, he became responsible for logistics [in the main production plant] at Oulianovsk. It truly exists between juridical entities, business units. Everything is possible in terms of mobility"*. She went on: *"I would say that, without people from workfloors [...], out of about three hundreds collaborators, I have at least fifteen in mobility at the moment."* (HR, Russian subsidiary).

In addition, training is an important part of HR policies in subsidiaries. *"We have a large training offer in Russia. We have training in selling techniques, time management, priorities management, how to do a presentation, etc. We also have training that are group oriented: training on Legrand's products, selling arguments for Legrand's products, marketing tailored for our products (practices and products of the group). [In 2010], the budget for training employees was 3 % of aggregate employment earnings (about 6 million rubles)"* (HR, Russian subsidiary).

Moreover, each subsidiary can use and taylor its intranet. *"There is an intranet at the group level, and at divisions and countries levels. They are translated in Chinese, Russian, English, etc."* (Edith Dumas, Head of Internal communication department). For these intranets, *"the base is hosted in France, frames, design are standardized for the group, and the content was made by myself and another person. It was Russian only."* (HR, Russian subsidiary).

Thus, subsidiaries appear as entities with clear boundaries and important within knowledge flows. As is well known from works on social networks, this kind of structure contributes to the creation of trust, identity, mutual understanding and shared knowledge within subsidiaries (Fang et al. 2010). This contributes to greater efficiency in that it helps subsidiaries improving their internal routines and being more reactive to their local surroundings. Simultaneously, the existence of a boundary contributes to filtering knowledge coming from the rest of the group (Zander and Kogut 1995). Although this may be detrimental, it nonetheless has the benefits of avoiding the introduction of inappropriate knowledge in a given subsidiary.

Summary of the "Downward" Stream

The phase we called "downward" stream can be characterized as a descending gradient of exploration from center to periphery. The phase starts in central R&D teams. At this level, significant exploration is conducted as new prototypes are built and industrial scenario are elaborated. These new products are then passed on what the firm calls "downstream" development teams located in different regions of the world. These teams take in charge the refinement and completion of the new products. At this stage, exploration still exists, but at a much lower levels than in the R&D teams, since most of the uncertainty has been removed. These new products are then put into production in the most convenient plants, depending on contingencies such as tax or logistics issues. Uncertainty is completely removed at this stage and the firm enters fully in an exploitation phase.

Three important elements contribute to the success of this gradual move from exploration to exploitation. First, the firm is endowed with a strong, common infrastructure (constituted of a matrix structure, platforms for modular products development, and information systems) that guarantees both a common knowledge and a certain routinization of the process. This helps in easing communications. Second, the firm displays a dense grid of intra-organizational linkages. These linkages can be formal or informal, directed or undirected and is sustained by a human resources policy that contributes to the development of the overall firm's social capital (Nahapiet and Ghoshal 1998). These interrelationships greatly help in moving from one stage to the other during the "downward" stream. Third, there are still some controls on the appropriateness of the innovation. In particular, subsidiaries may refuse to enter the production phase of a product that is not deemed appropriate for their specific markets. This ultimate filter limits the risk of undue investments (although it may lead to missed opportunities).

Role of IS in the Downward Stream

In the downward stream, the integration part of the ambidexterity process, we have been able to identify the role of various systems.

In early phases of the integration process, R&D departments and local development teams have to interact a lot to finalize new products and turn them into a form suitable for mass production. During this phase, since knowledge is not yet stabilized, supporting information systems must be such that they support flexibility and change. During that phase, accordingly, these are thus collaborative web 2.0-like tools that are rather used.

Later on, what is important in the integration phase is to reduce the differences between knowledge stemming from exploitation and knowledge stemming from exploration. Different information systems help in achieving this objective. Generally speaking, electronic communication channels contribute to lessen intraorganizational boundaries and facilitate communications among disparate units (De Sanctis and Monge 1999). This communication infrastructure contributes to the creation of a common context in which knowledge might be easier to transfer.

An important component of the information system is the intranet. The intranet is global in the sense that, although there are local adaptations, its overall architecture and a significant part of its content is framed and steered by the internal communication department of the headquarter. The intranet is used in wide variety of ways: "*Best practices will circulate quiet a lot on the intranet. If it's not secret or patentable, we will diffuse it through our magazine that is more widely forecasted. If we target people from engineering and design departments (new technology related to their job), then, we will maybe do a webcast to people from that profession, saying that's the new stuff, and here's how it works. If it is a new tool that facilitates the implementation of something, we will use the intranet, broadcast a demonstration video… It also depends on the perimeter of that new knowledge. It might be something of interest only for a site or a subsidiary. Each location has its own site, and we will make a link on it to diffuse it at the international level. […] For the R&D, there is SharePoint, it's a disposal for engineers. There are some SharePoint on some intranets. It's for projects and it replaces Notes databases [used to keep tracks of communications in the course of a project]*" (Internal communication department).

Although it may be the case that the intranet is also helpful in the upward stream for idea generation, respondents mostly saw it as a means to create a common knowledge base throughout the organization. The intranet is thus used for a great variety of purposes in a great variety of contexts. It appears as a flexible, though uniform infrastructure on which many forms of knowledge transfer can take place. It is a means to flexibly create links between different loci of knowledge production and use (Magnusson 2004). It enables both to standardize and personalize the exchanges and enables both local and global knowledge diffusion (De Sanctis and Monge 1999). The intranet is therefore useful for combining different forms of knowledge and can serve in combining old and new knowledge. In this regard, it is thus a useful support to integration.

5 Theoretical Framework and Propositions

5.1 An Overall Cycle

To date, the bulk of the literature on ambidexterity presents static ambidextrous settings (Raisch et al. 2009). The focus has been largely on structural design that could enable ambidexterity in organizations (O'Reilly and Tushman 2004; Gibson and Birkinshaw 2004). Yet, several studies point to the importance of time in developing new products, processes or capabilities within the firm (Brown and Eisenhardt 1997; Nickerson and Zenger 2002; Siggelkow and Levinthal 2003; Westerman et al. 2006).

Taking time into account in the ambidexterity phenomenon contributes to integrate the notion of ambidexterity in a systemic view of organizations (Martin and Eisenhardt 2010; Schreyögg and Kliesch-Eberl 2007). More precisely, it allows seeing firms balancing exploration and exploitation as complex adaptive systems that undergo continuous emergence of newness. The system is then modified by this newness that sparks modifications in its processes and cognitive structure. The firm then appears as a system that continuously reorganizes itself by balancing exploration and exploitation at different stages and in different ways.

Introducing time as an explanatory variable of a firm's ambidexterity, our case study proposes a process view on ambidexterity. It reveals that balancing exploration and exploitation is a process in which the entire organization is involved. We presented this process as made of an upward stream and a downward stream. In the upward stream, ideas are first generated by individuals at every level of the organization. These ideas are funneled to some specific units that assess them and retain the most promising ones. Retained ideas are then refined, elaborated and turned into exploratory projects. These projects are then submitted to the top management which selects the best ones or the ones that best fit in the firm's overall strategy. With the projects selected by the top management, the process enters its downward phase. Prototypes elaborated by research teams are passed to decentralized development teams that complete the new products and give them their final shape and functionalities. Once products fully elaborated, they are then transmitted to production units in charge of mass production.

The study also contributes to the concept of differentiation versus integration in the context of ambidexterity (Raisch et al. 2009). The upward stream, by allowing the transformation of ideas that emerge from daily operations into projects conducted in dedicated explorative entities is a differentiating process, in that it separates explorative from exploitative activities. Conversely, the downward stream, by bringing innovation back to operational levels is an integrating process in that it recombines old and new knowledge.

In the remaining of this section, we will further elaborate the different components of the identified cycle and discuss how it relates to the existing literature. Doing so will allow us to set propositions regarding the balance of exploration and exploitation in large firms.

5.2 The Upward Stream

In discussing the upward stream, we make two specific points. First, we propose that the decision dimension of this process is similar to a real option reasoning. Second, we point to the fact that exploration and exploitation are balanced not at one uniform level but that it is a multi-level process.

5.2.1 A Real Option Reasoning

Real option reasoning is the idea that the logic applying to financial option could be applied for managing other kinds of assets as well. A firm using a real option approach pays a premium to launch a project that may be maintained or abandoned at later point in time. Abandonment only entails the loss of the premium, not of the full-fledged investment. Hence, real option reasoning appears as a heuristic that endow firms with more flexibility as it maintains a number of possible development paths of future development at a limited cost. Bowman and Hurry (1993) popularized the idea that real option is a valuable heuristic for decision-making in organizations. In particular, several authors (e.g. Childs and Triantis 1999) argued that real option reasoning was particularly apt in managing R&D projects portfolio.

At Legrand, the upward stream can be considered as part of a real options reasoning mechanism (Hill and Rothaermel 2003). Agents generate elements of knowledge that convey the potential for future opportunities. Once a new development path is identified, the proposal is formalized so as to be considered by the decision makers and then truly becomes a real option for the organization (Barnett 2008). Strategists will then decide to exercise the most promising options (Maritan 2001). Exploration is then a proactive process unfolding from the generation of ideas to the decision to exert some of the available options by decision-makers.

One of the key issues in adopting real option reasoning is the option formation (Barnett 2008). At Legrand, ideas generation stems from every organizational level. It constitutes an additional source of new ideas and knowledge from which new business opportunities can be defined. However, for this new knowledge to become new business activities, it has to be assessed and validated by the top management. It is thus important that this knowledge finds their ways to decision-makers (Gibson and Birkinshaw 2004). Actors of contextual exploration must then find means through which to send proposals to top-managers and convince them of the value of their proposals (Dutton and Ashford 1993). To realize this, knowledge stemming from contextual exploration must be of a specific kind, and diffused through proper channels.

The success of a real option heuristic thus depends on the mastering of communication channels through to decision makers (Dutton and Ashford 1993). It does not suffice to know the administrative channels, but also to develop interpersonal relationships through which ideas will be channeled. At Legrand, informal networks appear to be very thick and new ideas seem to flow freely. This informal network is complemented by a process of progressive formalization of ideas. Burgelman (1991) stresses that the existence of such channels is a necessary condition for autonomous strategic proposals help in refining and adapting the induced, authoritative strategic orientation.

Moreover, the formation of real option entails to resort to different forms of ambidexterity at different stages of the process. We find evidence of contextual and structural ambidexterity, depending on the stage of development (Raisch 2008; Raisch et al. 2009). During idea formation, the ambidexterity is mostly contextual (Gibson and Birkinshaw 2004). Explorative activities (i.e. ideas generation) are carried out in the course of daily activities, while being encouraged and supported by a favorable managerial context (i.e. proposing new ideas is encouraged and valued). As ideas gain strength and are progressively turned into R&D projects, the ambidexterity becomes increasingly structural (Tushman and O'Reilly 2003). People in charge of developing these new ideas are more specialized, organized in dedicated teams and clearly identified work structure. Andriopoulos and Lewis (2009) note the importance for the success of an innovative firm to mix contextual and structural ambidexterity. Our study tends to confirm their results, while adding temporality: contextual and structural ambidexterities are not used simultaneously, but sequentially in a process of real-option generation. In this process, exploration and exploitation are increasingly differentiated (Raisch et al. 2009).

Prior research has identified important factors likely to play a role in the effectiveness of a real option approach for decision-making (Adner and Levinthal 2004; Barnett 2008). We expand these previous works by comparing the performances attained by various combinations of criteria. This allows us to highlight the fact that a complete understanding of a real-option based reasoning must take into account the interactions between different organizational features of the decision-making process.

However, it is likely that specific environments call for specific combinations of criteria. There should thus not be "one size fit all" real option heuristic. Instead, firms should devise specific ways to implement real option based reasoning, depending on the degree of the uncertainty characterizing the environment.

Proposition 1a: applying real option heuristic enhance the upward stream mechanisms.

Proposition 1b: setting proper channels and monitoring the appropriate type of ambidexterity along the upward stream will contribute to setting a real option reasoning heuristic.

5.2.2 Multi-Level Ambidexterity

Another interesting aspect unveiled by our study is the idea that exploration at one level may be compensated by exploitation at another level and vice versa. At Legrand, according to our respondents, the operational level, in subsidiaries is rather focused on exploitation, i.e. their primary driver is the achievement of production and sales objectives. In terms of time allocation, much more is spent to the realization of the prescribed tasks than on the generation of new ideas. At Middle management level, in particular at the headquarter, there is a much greater stress on exploration in that they are explicitly required to generate innovation (product innovation in divisions and process and managerial innovations in support functions). At the top management level, individuals pay great attention to the financial risk entails by novel projects and can thus be characterized as being rather exploitative.

Our results indicate that in using a real option reasoning heuristics for balancing exploration and exploitation in an R&D portfolio, firms have to find different balances and complementarities between the various mechanisms. Strategic and operational mechanisms should not be considered in isolation but in interaction with one another (Andriopoulos and Lewis 2009). Lacks and gaps due to organizational settings chosen at one level may be supplemented by settings at the other level. Hence, we can hypothesize that the balance between exploration and exploitation is not located at one specific level, but depends on multi-level interactions, one level compensating for the imbalance of others.

In this respect, our findings indicate that ambidexterity can be attained by the interplay between the different organizational levels (Raisch et al. 2009; Kang and Snell 2010). Our results indicate that drawbacks attached to exploration at the strategic level can be counter-balanced by exploitation at the organizational level and vice versa. So while the bulk of the literature addresses the issue of solving the exploration/exploitation tensions within one single level, either operational or strategic (Andriopoulos and Lewis 2009; Raisch et al. 2009), we point to another possibility, namely that of a multi-level ambidexterity.

Proposition 2: Within a firm, ambidexterity can be achieved in a multi-level way.

5.2.3 Implications for Information Systems

The first phase of the project is idea generation. In this phase, it is thus important to provide collaborators with tools enabling them to easily and casually submit new ideas. In addition, such tools would offer the possibility to centralize in a unique system all the ideas and proposals generated. Such tools already exist and we saw that Legrand started to implement them, relying mostly on Sharepoint.

However, in the upward stream, the true bottleneck occurs when funneling ideas to R&D teams. Given that there are relatively few R&D teams, but numerous ideas, there is a significant risk of missing opportunities or leaving good ideas unexploited, "idea overload" (Lindic et al. 2011). Moreover, this process is highly informal and do not seem to have been subject to a rationalization, while this could actually strengthen the flow of innovative ideas. Indeed, creativity and variability may be enhanced by providing specific rules, discipline and routines in the production of novelty (Farjoun 2010). At the same time, however, too much formalization would entail the risk of reducing knowledge heterogeneity, and thereby the overall firm's creativity (Kane and Alavi 2007). The management of this phase of funneling could thus be supported by a systematization of ideas gathering and channeling through to R&D teams, but attention should be paid not to stifle too much the process. It could thus be useful to have systems with the following functionalities: folksonomy, ranking, mapping of ideas, etc. Such systems would much more of an aid than on a control system.

Regarding ambidexterity per se, that is the question of defining the appropriate balance between exploration and exploitation, one has to distinguish between phases in which contextual ambidexterity prevails, from those in which structural ambidexterity prevails. When ambidexterity is contextual, it seems that the appropriate balance between exploration and exploitation must be left to individuals. In such cases then, the firm can equip people with software rather exploitation oriented and software exploration oriented, but let them decide on the amount of time they will spend in each type of activity.

When structural ambidexterity prevails, the management is more involved in that it has to decide on the appropriate balance between explorative and exploitive projects in the firm's technologies portfolio. The analogy we made with real options and options portfolio can inspire a proposal for a kind of decision support system that would help in striking the balance between exploration and exploitation at the firm level. One could imagine that a decision support system, relying on a real-option heuristic could help in selecting projects and managing the portfolio. Such systems already exist in other industry (e.g. gas or petrol industry) and could be adapted to the specificity of R&D projects management. Such a system should be able to value the firm's technological assets, to give an estimation of uncertainty surrounding projects and future cash flows. Estimation from past projects, experts' opinions, economic analysis of markets can provide information to do these estimations. It would then be possible to compare the different projects and consolidate them in a portfolio value. Such software could be a decision support system in balancing exploration and exploitation and mitigating risks.

5.3 The Downward Stream

While the upward stream is a differentiating process by which explorative activities are progressively separated from exploitative ones, the downward stream follows the opposite direction. Once new products have been prototyped and tested, they are re-inserted in the normal processes of the organization. The downward stream is thus an integrating process by which new and old knowledge are combined (Raisch et al. 2009). In what follows, we stress two specific points: the particular learning mechanism at work and the organizational characteristics that make this process possible.

5.3.1 Transformative Learning

From a cognitive standpoint, the process by which new and old knowledge is blended has been called transformative learning (Lane et al. 2006; Zahra and George 2002; Garud and Nayyar 1994). Transformative learning (Zahra and George 2002; Garud and Nayyar 1994): "*Transformation denotes a firm's capability to develop and refine the routines that facilitate combining existing knowledge and the newly acquired and assimilated knowledge*" (Zahra and George 2002, p. 190).

This process is essential as it is the link between exploration and exploitation (Lane et al. 2006). Thus, if the firm is to reap the benefits of its exploration efforts, the transformation process must be effective (Turner and Makhija 2006).

However, in the literature, transformative learning is often either overlooked or seen as costless and effortless (Turner and Makhija 2006). Our study reveals that, to the contrary, firms must devote a lot of attention and resources to this process. At Legrand, transformation occurs in a stepwise manner. First, prototypes and plans are elaborated in R&D teams. The output of this first phase remains largely exploratory knowledge, far from operational considerations. This output is then passed on to development teams in charge of developing further the product and bringing him closer to an implementable form. Next, the project passed on production plants where it becomes a fully exploitable product. Thus, far from being a natural diffusion process, the transfer of new knowledge through to the units that should use it is a carefully monitored process (Turner and Makhija 2006).

Moreover, the interpretation of this new knowledge by agents who will eventually use it is progressive. New knowledge goes through different airlocks in which interpretation of the new product is progressively built. The process thus takes the form of a descending gradient in terms of explorative degree of the knowledge transformed. As new knowledge goes down the process, uncertainty attached to it is progressively reduced and answered are brought to any question posed by the actual mass production and commercialization of the new product: costs and investment involved, functionalities that have to be added, production tools that have to be implemented, documentation that have to be written, etc. Beside uncertainty reduction, the process also mixes old and new knowledge. Usually, the firm relies on existing industrial equipment to produce new products. Hence, people from production have to recombine their existing routines and practice to incorporate newness into known and mastered processes.

Proposition 3: Adopting a stepwise integration facilitates transfer and interpretation of new knowledge.

5.3.2 A Matrix Structure

This process is made possible by the organizational structure adopted by the firm. The organizational structure embeds communication and coordination mechanisms that ease knowledge flows within the firm (Taylor and Helfat 2009). In the case of Legrand, the structure is a matrix.

In terms of transformation, knowledge will flow through divisions. The first benefit of this is that all agents belonging to the same division share a minimum common knowledge about products. This common knowledge base facilitates exchanges between the different entities constituting the division. In addition, within a division, it is easier for middle managers to maintain communication networks and informal linkages between the different constituents. There are numerous exchanges between the R&D teams, the development teams and production teams throughout the transformation process. These exchanges greatly facilitate transformation. More generally, social integration mechanisms, both formal and informal that can be deployed within divisions contribute to the transformation of explorative into exploitive knowledge (Zahra and George 2002).

The matrix structure presents another characteristic of interest for transformation. One may also note that the matrix structure allows implementing various controls on the transformation process (Cardinal 2001; Turner and Makhija 2006). First, there are controls on outcome of the R&D process. This control is made by middle managers heading countries. Because of dual responsibilities, middle managers in charge of subsidiaries can veto some innovations (Siggelkow and Rivkin 2005). Matrix structures thus offer another control levels that may prevent the introduction of new knowledge that could potentially harm the performance of a given subsidiary. One can expect that, given the general objectives set for these managers, they will tend to abandon what they perceive as falsely promising. There is however the risk to reject false poor projects (Adner and Levinthal 2004). This risk is nonetheless limited by the fact that a head of a country is but one actor along the decision chain and that is decision is mitigated by other decision makers (Christensen and Knudsen 2010).

In addition to these outcome controls, the firm is also likely to use clan controls in the transformation process (Ouchi 1980). Clan controls are social controls exerted by an entity. Newness is then confronted to existing norms, values and shared practice prior to be accepted. At Legrand, given strong local identities and shared knowledge, it is likely that new knowledge have to be first assessed and interpreted by the local subsidiary as a whole prior to be integrated in the set of practices. This additionally contribute to the robustness of the transformation process and guarantees that new knowledge would fit the specific local subenvironment of the subsidiary.

Proposition 4: Adopting a matrix structure facilitates transformative learning.

5.3.3 Implications for Information Systems

In this downward stream, information systems could be useful in supporting the transformative learning process. This process is made of two principal components: knowledge transfer and knowledge interpretation (Turner and Makhija 2006). Regarding the transfer aspect, an information system should help in translating knowledge into representations understandable by receivers (here, people from subsidiaries). Such a system should help in depicting and reflecting upon an understanding of the new knowledge arriving at the subsidiary. It would thus not be a simple transfer of codified knowledge but a tool that would help in transforming knowledge from one form into another.

Regarding interpretation, a system with functionalities that help in making sense of newness and incorporate it into the existing knowledge base would leverage transformative learning. Such a system should enable people to relate arriving knowledge to their existing knowledge base. This should in turn allow them to modify the current knowledge base and sparks a reflection on their practices and ways of doing things, and how these practices are modified by new elements.

Intranet 2.0 seems to be suitable support for this kind of cognitive processes. Given the ability of such systems to aggregate different forms of knowledge, both stemming from exploration and from exploitation (in posts, blogs, wikis, but also more formalized documents and institutional pages), it is appropriate for agents who can access and combine different forms of knowledge to evolve their representations as newness is introduced in the production process.

6 Mathematical Modeling

6.1 Set of Hypotheses

In this section, we attempt to analyze the predominant factors we have extracted from the case study to develop a mathematical model for explaining the interplay in managing ambidexterity. We will present below our findings via three mathematical models which related to the three different view on the functioning of innovation process as it emerges in a multinational company.

As we have stated above, the key objective of a MNC, is the efficient coordination on all levels and global coherence of the firm aiming on optimizing for costs and investments, gaining the market shares in the countries where it is present, assuring the increase in profitable sales and speeding up the new product development process.

It becomes apparent that one of the challenges for a MNC is to combine successfully three processes:

1. global coordination of the increasing number of subsidiaries on the international markets;
2. adaptation to the sub-environment of local markets (local responsiveness); and
3. continuous innovation.

A strong emphasis is given on future orientations of the firm via an acquisition strategy of local leading companies in countries or market segments with high growth potential.

Research on intra- and inter- organizational relations focused mainly on the efficient coordination which may be categorized into two types: centralized coordination and decentralized coordination. Here we will give emphasis on the centralized coordination witnessed at the headquarters' level in the multinational company which we have studied. However it is combined with the great autonomy lend to subsidiaries on the local level.

Centralized coordination involves a unique decision-maker, in our case, Division of General Managers of the Headquarters, who possess all information on the whole organization that is relevant to make decisions on innovations (at the headquarters' R&D labs) as well as the contractual power to implement such decisions (via the R&D teams). In this case, all the partners involved (subsidiaries, R&D project teams and headquarters) gain from the interactions and receive revenue according to their respective investment.

In this global approach the main objective is to maximize profit based on the acquiring a new subsidiary either to insure/strengthen the presence on a new market or to support the evolution of the core business of the organization. The main challenge of this coordination is to design the interrelations which maximize the organization's profit. The capacity to innovate and rapidly adapt to the new environment depends on the structural design of the company and on its managerial practices, and it effects the profit of both the subsidiary and the headquarters. This phenomenon explains the interdependencies of one over others and forces the partners towards cooperation. This cooperation is essential to stay and flourish in the business. So, to develop the interrelations where both sides are equally benefitted is a complicated issue. There are vast research efforts devoted to the interrelationship issues in a MNC between headquarters and its subsidiaries (e.g. Magnusson 2004; Fang et al. 2010; Lahiri 2010).

As is has been stated in the conceptual framework presented above, our study reveals that different forms of ambidexterity are present at different stages of the innovation cycle. Early in the upward phase we observe contextual ambidexterity during which new ideas are generated. In the downward phase, structural ambidexterity becomes apparent as exploitation is conducted mainly in the different subsidiaries. We have observed that ambidexterity is supported through specific structural (common standardized procedures put in place) and managerial features (such as internal mobility and informal networks which contribute into creative ecosystem.

We first briefly introduce different hypotheses that determined our modeling choices concerning intra firm collaborations within a multinational organization, between headquarters and subsidiaries as well as among subsidiaries. We focus our attention on the effects of the knowledge and information exchange. The effectiveness of a bilateral collaboration is determined by cognitive, relational and structural embeddedness.

In many models of joint innovation, a simple measure of distance in knowledge space determines the effectiveness of collaboration (see for example, Nooteboom 2000, Cowan et al. 2007; Peretto and Smulders 2002; Mowery et al. 1998). We insist on the fact that putting two subsidiaries or a headquarters' and a subsidiary's knowledge together requires considering deeper issues of complementarities,

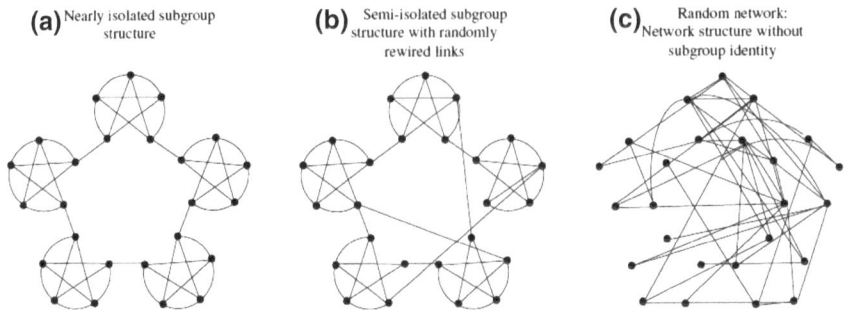

Fig. 2 Three types of organization structures. Fang et al. (2010)

beyond the specification of an optimal distance as they are involved in exploitation activities.

Balancing between exploration and exploitation is in the focus of another stream of research models based on the solutions obtained on an NK-landscape (Siggelkow and Levinthal 2003; Siggelkow and Rivkin 2005) It has been shown that the value of decentralization and the optimal amount of integration is contingent upon both the amount of environmental turbulence and complexity faced by a firm.

Feng et al. (2010) studied the three different types of organization structure: nearly isolated subgroups; semi-isolated sub groups and random network (see Fig. 2). They constructed a generalized payoff function and run a series of simulations to examine the impact of inter-relational network structure on organizational-level performance. Their findings show that parallel isolated learning within each subgroup enables exploration, and the preservation of variety of knowledge in an organization while learning across subgroups enables exploitation by facilitating the rapid diffusion and assimilation of currently superior knowledge. Therefore, a productive balance between exploration and exploitation in a MNC can be achieved by giving more autonomy to subsidiaries with a small fraction of cross- group links, in other words with some central subsidiaries of intermediate level possessing the R&D project teams

Our first hypothesis relates to the structural characteristics in a multinational firm with regard to exploration and exploitation activities. Additional hypotheses deal with managers' roles and informal relationships within subsidiaries. During the upward phase ideas are sent from operational to middle management levels, and after being explored more in depth and formalized into possible projects, middle management submit these ideas to top management. Top management then decides, in a rather conservative way, which projects need to be financed, the investment strategy is developed. This procedure could be considered as projects' portfolio management. A downward stream is moving from exploration at the headquarters to exploitation in subsidiaries.

So, we suppose a permanent link between differentiation and integration of new and old knowledge in a MNC. Two, possibly contradictory, actions have to be performed: first, ensure that new knowledge will flow through to subsidiaries. This step strongly depends on structural issues—the firm need to rationalize, standardize procedures to ease knowledge communication. Second, ensure fit between subsidiaries and their local environment, which supposes that a subsidiary should be autonomic on a local level.

The overall logic of our model, which encompasses the hypotheses stated above, is as follow. Firms have to find a balance between exploration and exploitation activities: they earn money merely from exploitation activities in subsidiaries though they need to engage in exploration activity in order to expand their market and assure continuous innovation process. To decide what exploration moves should be generated and dedicate resources to either exploration or exploitation, top managers have to get a perception of their environment and communication downward and upward facilitates this process.

6.2 Mathematical Models

We assume a finite population of subsidiaries $S = \{s_1, s_6, ... s_N\}$, each subsidiary s_i is a function of two variables $S_i = F(L_i, R_i)$, which describe exploitation, L_i, and exploration, R_i, activities respectively. In other words we could refer to these variables as to old and new knowledge in firm's possession. Each subsidiary is characterized by :

– the benefits a subsidiary gains via exploitation activities L_i;
– amounts of K distinct types of new knowledge (innovations) it holds K_i, (the types of knowledge could certainly differ in different subsidiaries, while headquarters possess the complete information on all these distinct types of new knowledge);
– incertitude associated with each innovation σ_i.

There are two possible types of subsidiaries—those of the intermediate level where R&D project teams develop some exploration activities $R_i \neq 0$, while other subsidiaries pursue only exploitation activities L_i and $R_i = 0$, so $S_i = F(L_i, R_i) = S_i = F(L_i)$, $i = 1,...N$ By innovating a subsidiary creates new knowledge and so moves into a new location in the knowledge space.

It is important to stress that innovation is knowledge creation—whether it is an R&D project pursued by an intermediate subsidiary or by a new technology elaborated by an R&D lab of the headquarters. In both case it is supposed that knowledge creation leads to product innovation, the innovation is put in place, a firm gains in knowledge and the involved subsidiaries line up their profiles which become then similar $S_i = F(L_i, R_i) = S_j = F(L_j, R_j)$. Simulation brings us to this type of dependence within a limited time frame (Fig. 3):

Fig. 3 The impact of
innovations on knowledge
evolution in subsidiaries
(a case of 5 subsidiaries)

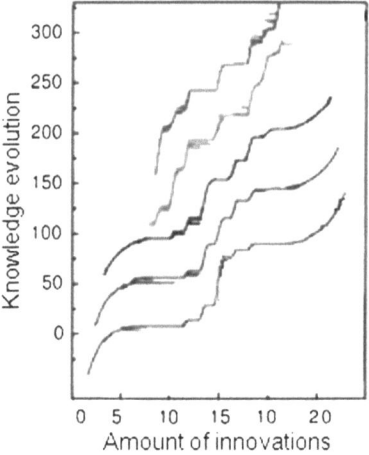

6.2.1 The Knowledge Evolution Approach

To be in line with our hypothesis, we consider that the collaboration of the R&D team project on an intermediate level brings benefits to all subsidiaries having relationship to this one and creates knowledge flow vector. Formally the creation can be described as

$$dF^t\left(L_i, R_i\right) \ = \ \lambda_I R_I dt \ + \delta_i * dW_i^t;$$

here λ_I represents new knowledge evolution in time, which corresponds to the growth of exploration activities; we describe the variation of knowledge acquired by the subsidiary. The learning process which the firm will realize within the process focuses on the estimation of $\lambda_I R_I$ despite the presence of noise—δ_i: the higher is the value of δ_I, the more difficult the knowledge evolution process is, the higher are the expenses; dW_i^t is a Wiener's process growth of stochastic nature which follows a normal distribution law in a time limit of one R&D project life cycle.

We suppose that at the initial stage a firm has a natural tendency to exploit and as hypothesized above, a firm earns money only from their exploitation activities (L) while the knowledge vector is described by $F^1(L_i, 0)$. However, exploitation activity exhausts the market, and at some point firms need to renew it. They then draw on exploration activity (R) in order to expand their market. We see exploration as an activity through which an organization is able to renew its processes and products, and hence expand its market. These two types of activities, exploitation and exploration, can certainly be developed in parallel. Formally, this is exploration which will move the organization from one state in which the market is saturated to another, unsaturated state, in which the firm can expect higher profits $F^2(L_i, 0)$—see Fig. 4.

We assume certain expenditures associated with each R&D project during the exploration phase (R). As soon as the R&D project is accomplished (at the end of the

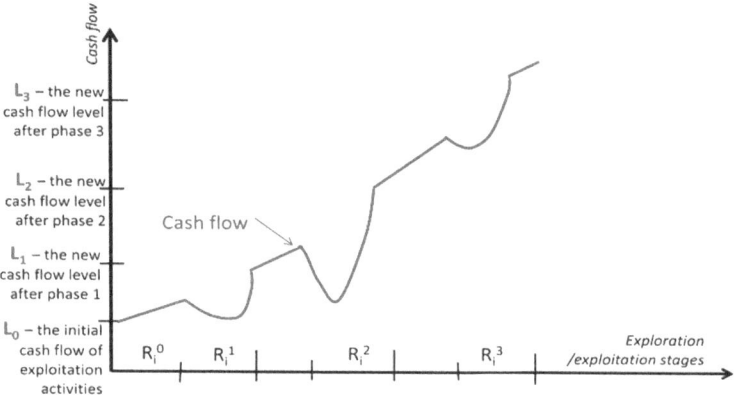

Fig. 4 The cash flow differences of a MNC in the continuous innovation process

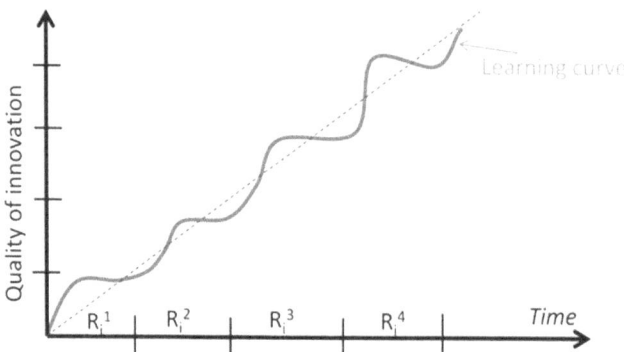

Fig. 5 Firm learning curve of the four exploration phases in the continuous innovation process pursued by a MNC

phase R_i^1, for example), the innovation strategy is transferred into an action plan for new exploitation activities in headquarters and subsidiaries, and the cash flow of exploitation activities of a firm increases; a firm moves to a new level. There is thus a trade-off between the benefits gained from exploitation activities and the investments to be made from exploration at the beginning of a new phase (R_i^2).

Further on we consider that there exist some R&D projects that are more interesting than others in terms of new knowledge creation (exploration process of R_i activities). We suppose that $F(*,R_i)$ is a concave or at least a quasi-concave function. The example of such a function defined on a limited time period t (e.g. the phase R_I^1 or R_I^2) is : $F(L_i, R_i) = L_I - (a_i R_i - \beta_i)^2/e^{Ri}$

In other words, the exploration process has a limited duration period and a firm moves into exploration on a new level of acquired knowledge $F_i^t = F^t(L_i, 0)$ as

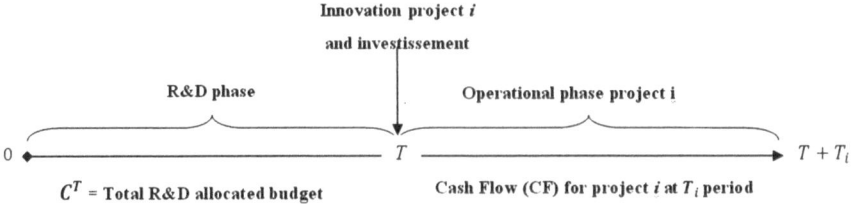

Fig. 6 Analysis of innovation projects and a choice of a best project

soon as exploration activities are closed and the R&D project is accomplished. This happens when subsidiaries on a lower level increase their L_i activities.

The inter- and intra- relations between subsidiaries contribute into the increase of a knowledge production function. For simplicity, we use (following Cowan et al. 2007) a constant elasticity of substitution function, so the new knowledge created by the R&D project team is

$$\Phi\left(s_{i,}s_{j}\right) = \alpha\left(\sum_{i=1}^{N}\left(F_i^t\right)\right)^{\frac{1}{\gamma}}$$

If the innovation process is successful new knowledge is created and added to a firm's portfolio of knowledge. It seems natural to let this point be an inflexion point (see Fig. 5 as an illustration).

In what follows we attempt to capture these characteristics applying first, the cash flow and the production approach, second, the knowledge evolution approach. We focus on organization design model which sums up our findings on the effectiveness of knowledge penetration and knowledge transfer among the head-quarters and subsidiaries of a multinational company.

6.2.2 The Cash Flow and the Production Approach

The knowledge flow approach focuses on the best choice of a project for financial support. We assume that a firm develops the R&D process for a limited time period, till $t = T$ and then decide either to invest or to withdraw. To make a final decision a MNC will compare the expected values of net benefits of the considered projects. The project that will be chosen at time T can be different and will correspond to the R&D strategy of a firm. Then the decision making process can be described through the following stages (Fig. 6):

For each project at time $t > T$ we could make *cash flow* estimation $CF_i^t = E\left\{\left(\sum_{t=T}^{T+T_i} F_i^t\right)\right\}$.

If a firm invests at date T in a project i, then the expected net value of a project by date T is

$$E(VN_i) = E\left\{\left(\sum_{t=T}^{T+T_i} F_i^t\right)\right\} - C^T$$

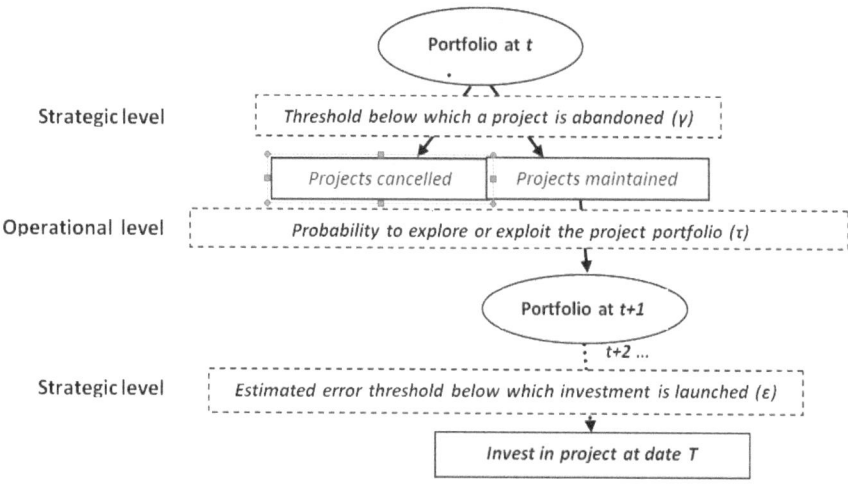

Fig. 7 The final decision making process of a firm and the key criteria to control the process

If learning process allow to reduce uncertainty about F_i^t by the date T, then $VN_i = \sum_{t=T}^{T+T_i} F_i^t - C^T$.

We assume that a MNC operates in the environment of several innovations considered at the same time (either suggested by the subsidiaries or developed by the R&D laboratory of the headquarters); and the objective is to determine the best innovation to invest into ("best" in terms of costs, technological equilibrium and feasibility). To make a choice, a firm employs a learning process and attempts to balance two processes—focusing of each particular idea, study it in details in order to reduce incertitude and risks involved or perform the most accurate possible evaluation for each project suggested in order not to miss a god opportunity. Different operational mechanisms could be implemented for that.

We suppose that a firm has a limited R&D budget and can't afford to pursue more than n technological projects out of m possible. At each moment t a firm should learn about one or another innovation opportunity (potential idea) and the cost of learning c is defined as for each innovation, so we get $C^T = \sum_{t=0}^{T} c^t$; the R&D costs correspond to the sum of total primes paid at the R&D phase.

As soon as a firm chooses one idea out of all possible (in the portfolio), a firm gets the period t, after paying a prime c, a quality evaluation CF_i^t of this innovation is made.

Within the time, a firm needs to precise better the estimations of different projects' quality via a learning process. L'entreprise affects a quality measure Q_i^t for each projet at time t. Q_i^0 is set up at 0 at the beginning. Every time a technologie i is chosen, the value Q_i^t corresponds to the expression below (following Sutton and Barto 1998),

$Q_i^{t+1} = Q_i^t + \alpha X[s_i^{t+1} - Q_i^t]$, with $0 < \alpha \leq 1$, and where: $s_i^{t+1} = F_i^{t+1} - c$.

In this process, α ia the weight of novelty for each project: the higher a is, more new elements of importance will be negotiated as these elements describe the importance of the environment.

In other words, α can be considered as a size of a firm: the smaller α is, the more it is affected by the expected value evaluation for each project in a long term. At the R&D phase, a firm observes the F_t^i data of the project on which a firm is learning; Q_i^t is recalculated, and estimate average *cash flow* λ_i^F generated at each stage of the considered process. $\sum_{t=T}^{T+T_i} E(\lambda_i^F)$, remains the total project i cash flow according to firm's estimations. Figure below illustrates the decision-making process that should be put in place by a MNC to choose the best innovation project to invest based on the cash flow criteria (Fig. 7).

6.2.3 Modeling Organization Design Using Josephson Effects

As it has been stated in part 3 of the report, a wide range of research on modeling of organization structures focused on the possibility of using organizational structure as a mechanism for managing the balance between exploration and exploitation and focused on the role of centralization/decentralization (e.g. Ishida and Ohta 2001; Nickerson and Zenger 2002; Siggelkow and Rivkin 2005; Miller et al. 2006; Taylor and Greve 2006; Fang et al. 2010). By decentralizing the learning process to subunits, in our case—to the subsidiaries of a MNC—and providing barriers to the rapid diffusion of ideas and norms across the subsidiaries, a MNC assures the exploration of a more diverse range of solutions which is encouraged by top management. It has been shown that gains might be significant in a case of isolating R&D teams from the mainstreams of organization (Bower and Christensen 1995; Benner and Tushman 2003; O'Reilly and Tushman 2004) due to exploring new alternatives unfettered by the demands and norms of the rest of organization. A classic example is the development of Macintosh design in the 1980s by Steve Jobs who did not believe in growing corporate environment at Apple (Rogers 2003).

Integrating the lines of organization structure and organizational learning reveal the factors as the degree of complexity and dynamism or interaction in the firm's internal environment—the degree of differences in between subunits involved. The question we pose—how to combine the differentiation of activities in a MNC and integration of new knowledge across the boundaries?

To address the problem we refer to the Josephson effect, well known as a the phenomenon of physics describing an electric current transiting across two structures, weakly coupled superconductors, separated by a very thin insulating barrier. The challenge is to adapt the scheme to the knowledge flow and the new ideas penetration between a headquarters and a subsidiary and among subsidiaries and describe this process by means of the physical model. We investigate the diffusion of knowledge across the barriers of a MNC and consider a transfusion of new knowledge which passes on both directions: from a subsidiary to headquarters and then from headquarters to a subsidiary.

The arrangement—two superconductors linked by a non-conducting barrier—can be seen as a headquarters and an intermediate subsidiary possessing different types of knowledge. One of the important characteristics of the Josephson effect situation is a "phase factor". Phase in physics of material science is defined as a region of a material that has the same composition and structure throughout, and is separated from the rest of the material by a distinct interface; moreover a phase may contain one or more components.

Referring to a MNC, we consider a phase factor as a specific distinct environment in which a headquarters or a subsidiary functions. In our view, cognitive, relational and structural characteristics define a phase factor and therefore will determine the effectiveness of a bilateral collaboration.

In physics it is known as a Josephson junction; the current that crosses the barrier and moves the "electrons" into another superconductor—the Josephson current- will bring new technological knowledge (innovations on the level of subsidiaries).

The basic equations governing the dynamics of the Josephson effect in physical model are

$$U(t) = h/2e * d\phi/d\tau;$$

this is a superconducting phase evolution equation which corresponds in our case to the knowledge evolution from headquarters to subsidiaries.

$$F^t(L; R) \sim I(t) = I_c \sin(\phi(t))$$

describes Josephson or weak-link current-phase relation and corresponds to the "interrelationship" knowledge flow between a headquarters and a subsidiary.

We note that $U(t)$ and $I(t)$ are the voltage (power of relations) and current (knowledge flow) across the Josephson junction, $\phi(t)$ is the "phase difference" across the junction (i.e., the difference in phase factor, of a headquarters and a subsidiary). I_c is a constant, the critical current of the junction, which in our case describes the initial knowledge flow between the units. The critical current is an important phenomenological parameter that can be affected by temperature (e.g. intensity of interrelations between a headquarters and a subsidiary) as well as by an applied magnetic field (e.g. specific investment constraints put in place). The physical constant $h/2e$ is the magnetic flux quantum, the inverse of which is the Josephson constant. In our understanding, it is a measure of potential differences between a headquarters and a subsidiary.

In practice the Josephson effect provides an exactly reproducible conversion between frequency and power of interrelations. Since the frequency can be defined precisely and practically by the MNC standards, the Josephson effect can be used to model for most practical purposes, the interrelations among a headquarters and a subsidiary.

7 Conclusions

We sought to answer the question "How a multinational corporations can manage to be ambidextrous?" To do so, we relied on a single case study carried out at Legrand, a highly successful multinational company.

Our main finding is that in order to answer our research question, one must take a process view on ambidexterity. In the studied company, ambidexterity is achieved through a process in which the entire organization is involved. It starts with idea generation from any place in the organization. These ideas are funneled towards specific explorative teams that further develop and refine them. These explorative projects are then submitted to the evaluation of the top management that manage the overall firm's technological portfolio. Agreed projects are then progressively transformed into exploitable products, following a stepwise transformative learning process.

The upward stream of this process is made possible by a real option reasoning process. This process allows the generation of options that can be exercised or abandoned by the organization, thereby providing it with flexibility in the early phases of exploration. The downward stream of this process relies heavily on the matrix structure of the organization that permits both to transfer knowledge along divisions from one point to another in the organization, while also introducing further check points controlled by middle managers that can assess the value of an innovation further downstream in the process.

The current report suffers some limitations. Most of all, due to difficulty to access our empirical field, we have only be able to access the Russian subsidiary. The empirical research is still ongoing and data are still to be collected on the Indian subsidiary. This latter part will be the object of a follow-up report.

Annexes

Annex 1: Interview Guidelines

1. **(Differentiation) Approach to Exploration/Exploitation?** Both or one of the two at HQ? Both in some subsidiaries and one of the two in other subsidiaries? If both everywhere: Partitioning/sequential/contextual?

   a. How do you go about researching new products, new markets, new countries etc.?

    b. Do you locate such marketing research, business research, and R&D in separate units, different from those engaging in more 'basic' research?

    c. Do you have the same units working on 'new' research and 'basic' research at separate points in time?

    d. Do you have separate project groups for each purpose? (If project groups : stability, tenure and duration, composition etc.)?

    e. Do you attempt to create units with capabilities to do both types of research simultaneously? Does this play a role in the subsidiary location decision? If yes, how so?

    f. How do you go about allocating managerial responsibilities for exploration and exploitation? What control systems do you use?

    (based on Simsek et al. 2009 and Kang and Snell 2009).

2. **Integration**

    a. **(Integration at the top management level) Leadership**, top management capabilities, founder orientations (including geocentric, polycentric, and ethno centric for MNC org.)

    Shared vision and development, clear articulation, commitment to exploration AND exploitation, regular meetings with key execs, TMT behavioral integration (ability to manage and handle contradiction), significant common and diverse experience of TMT, transactive memory of TMT

    Financing of radical innovation/incremental innovation. How frequent? What magnitude?

    b. **(Integration at operational/middle managers levels)**

    Encourage improvisation within projects?

    Encourage communication among projects (tram level or project manager level)? At the subsidiary level? At the firm level?

    Seek unconventional linkages among ideas within and among projects?

    Projects are technology-driven or customers-driven?

    NB: integration at the individual level are addressed in the "human capital" section

    Based on Andriopoulos and Lewis 2009.

3. **Knowledge Circulation**

    a. Explorative knowledge

    How do innovative ideas circulate within the firm? Is it at the site/national/ firm level?

    Do you know of the adoption/reuse of innovative ideas? According to you

what are the success/failure factors?
What are the means for innovative knowledge diffusion (meeting, IT, interpersonal networking, job rotation)?

b. Exploitative knowledge
How do best practices circulate within the firm?
How is valuable technologies diffusion handled?
Is it at the site/national/firm level?

4. **Performance**
Can you assess the impact of your knowledge management in terms of innovation generation on the firm/subsidiary's economic performance?

– Number of successful innovative products put on the market
Can you assess the impact of your knowledge management in terms of best practice circulation on the firm/subsidiary's economic performance?
– Time saving (time to market, project development length)

5. **Human Capital**
Ethnocentric/Polycentric/Geocentric Selection Practices; Use of Specialists versus Generalists; Nature of Performance Management; Mkt-based versus equity based compensation schemes; mkt-based versus ILM based employee relations; specialist versus generalist managerial development practices.

   a. How do you approach staffing at HQ and in international locations? To what extent do you prefer French executives versus locals or third nationals in each of these locations?
   b. To what extent do you get involved in selection of executives and/or employees in subsidiary locations? To what extent is this responsibility delegated to each of the subsidiaries? Are there any exceptions to the general rule? What about R&D for example? Or Marketing?
   c. To what extent are HR practices centralized and/or standardized across subsidiary locations?
   d. In this country, compared to other players in your industry, do you have broader job definitions or narrower job definitions (more generalists or more specialists in jobs)? Could you highlight a few key examples?
   e. In this country, do you offer/support broader and multiple skills development for your executives and other employees or do you offer/support within-job or within-function skill development?
   f. In this country, what is your approach to job rotation? Do you use it? Within specialized areas or across functional or business units?
   g. In this country, what is your approach to hierarchical movement (internal promotion) of executives and employees, vis-à-vis external recruitment of people into these positions?

    h. In this country, on what basis do you make selection decisions? Based on cognitive ability and aptitude testing on the one hand or specific job competencies and job knowledge on the other hand? What other selection mechanisms do you use?

    i. In this country, do you utilize skill-based or knowledge-based pay as part of your overall compensation system? Do you for example pay for ideas or for reputation?

    j. In this country, to what extent is your pay system based on market factors?

    k. In this country, to what extent do you use performance based pay as part of your overall compensation system?

    l. In this country, to what extent is seniority a key factor in your compensation system?

    m. How woud you describe your performance appraisal and management system? Is it focused more on behaviors and observation and rating of these behaviors or on results and outcomes? Could you share with us a sample of your rating scales?

    (above based on Kang and Snell 2009).

6. **New Product Development**

Organisation of the NPD; process versus product orientation; location of NPD or R&D (centralized, decentralized, distributed etc. in MNC operations); NPD links with IS; R&D cultures—convergent across subsidiaries or divergent.

    a. How do you handle demands for innovation that go beyond existing products and services?

    b. What is your intent and strategy for inventing new products and services?

    c. Do you experiment with new products and services in local and global markets? How? In what order/sequence?

    d. How do you go about commercializing products and services that are completely new to you?

    e. How do you take advantage of new opportunities in new markets?

    f. How do you handle the search for and adoption of new distribution channels?

    g. How do you handle the search for and acquisition of new clients in new markets?

    h. How do you handle entering new technology fields?

    a. How frequently do you refine the provision of existing products and services and how do you go about it?

    b. How do you go about small adaptations to existing products and services? With what frequency?

    c. How do you introduce improved but existing products? First in the local market and then in the global markets? Or the other way around? Similarly around the world? Or Differently?

d. How do you improve the provision (mfg. & selling) efficiency of your products and services?

e. Do you work on improving the economies of scale in existing markets? If so how?

f. How do you go about expanding the product/service offering to existing clients?

g. How do you value lowering of costs of internal processes? What do you actually do to attain this objective?

h. What, if anything, do you do to improve existing product quality?

i. What, if anything, do you do to improve production flexibility?

j. How do you go about reducing material consumption of increasing yield? (Above based on He and Wong 2004; Jansen et al. 2006)

7. **Organizational Capital**

HQ/Subsidiary links and relationships; Formal and informal networks within and across subsidiaries and with external entities; Matrix versus product team structures, nature of differentiation and integration, **tightly coupled subsidiaries with loosely coupled interrelationships** (or loosely coupled subsidiaries with tightly coupled interrelationships). (Subs in EMs: co-orientation, co-competence, co-opetition, co-evolution), mechanistic/organic nature.

a. How would you characterize your organizational structure across and including all international locations? Divisional, product-team, matrix?

b. To what extent are some of the key operations centralized or located in a few countries or regions? R&D, manufacturing, marketing?

c. What is your approach to integrating across subsidiaries? None (leave them alone and independent), moderate, or high (constant and frequent interaction among subs and with HQ)?

d. How would you characterize the organizational culture that exists in your organization (here at HQ, here in India/Russia)? Would it be more along the lines of conformity to established rules, norms, and procedures or more along the lines of encouraging people to proactively create, shape and respond to established cultural values and norms?

e. How would you describe the way your organization functions?

f. In terms of standardization of procedures and structures?

g. In terms of desired patterns of behaviors (establishing ingrained patterns or creating capability to enact and respond to challenges)?

h. Centralization versus decentralization?

i. Many rules or few rules?

j. Many teams or few teams?

k. With subsidiaries in Russia and India in particular, to what extent do these following factors play a role? What is your approach in each case?

1. Exploit existing assets for short term profits versus acquiring new assets for long term growth?
2. Deploy, exploit, and utilize transactional (mkt-based) capabilities versus relational (network-based) capabilities of the subsidiary and its key managers?
3. Does your subsidiary simultaneously compete and cooperate with local rivals, local partners, local suppliers, distributors, and governments? How so?
4. Does your subsidiary simultaneously engage in complying and responding to local rules, customs, and laws, while also attempting to influence and change them?
   (based on Simsek et al. 2009; Kang and Snell 2010)

8. **Information Systems**
   IS architectural features, IS—OC links; IS/NPD links; IS/network links

   a. How do you approach linking the HQ with all the subsidiary operations? What IT tools are used to link them all? Or do you leave them all as standalone operations?
   b. Do you make available KRs and databases from HQ to all the subsidiaries, specific tools to specific subsidiaries, on a need to know basis, some other arrangement, etc.?
   c. Do you use different KRs at the HQ and each of the subsidiaries or one large KR with complete access/multi-level privileged access to different subsidiaries and units?
   d. For R&D, do you have an ECOP that is predominantly global or local, situated in each unit, with possible links to the other ECOPs in other locations?
   e. Global/local Intranets? Are ECOPs accessible through the intranet?
   f. Can all project teams access database/intranets/collaborative spaces in which innovative ideas/proposals are stored/worked on?
   g. Conversely, can teams working on breakthrough projects consult the different databases/use the different tools used in more routinized activities?
   h. What kinds of IT networks with customers do you use, if at all? What are the benefits derived?
   i. What kinds of IT networks with suppliers, distributors, and logistics partners, do you use, if at all? What are the benefits derived?
   j. Does your ERP system provide for connectivity to customers and suppliers and other partners? How so? What benefits, if any?
   k. How does your IT network coordinate with the new product development process?
   l. With the new business/market development process?
   m. Do you provide for the creation of IT based networks for evolving interest groups (formal or informal)?

Annex 2: Map of Our Key Findings

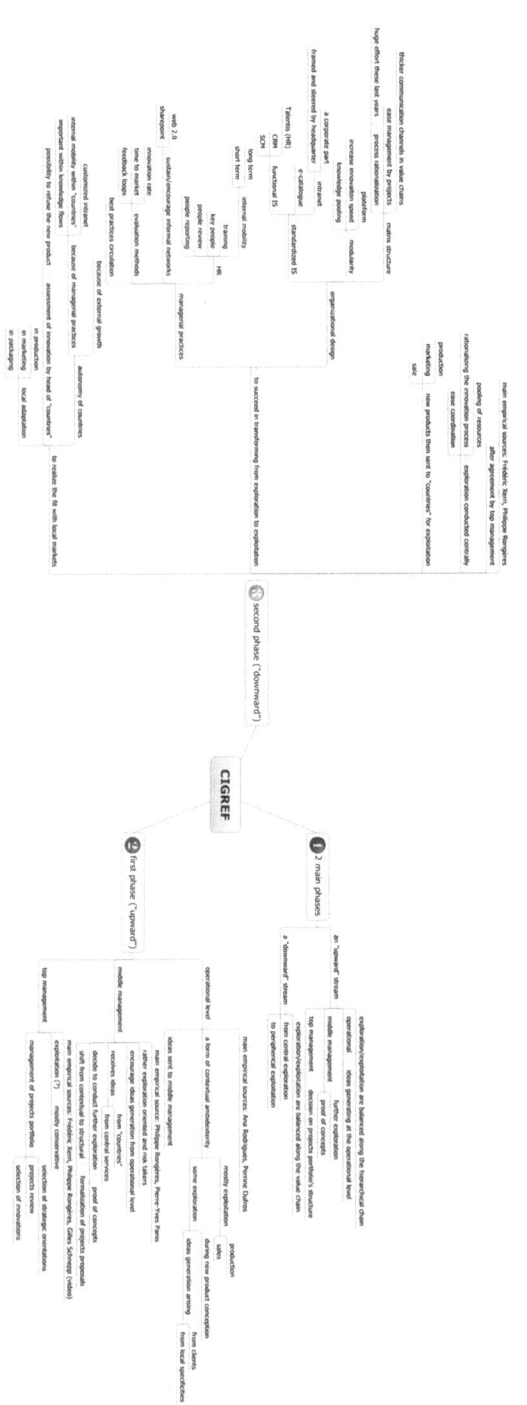

References

Adler PS, Goldoftas B, Levine DI (1999) Flexibility versus efficiency: A case study of model changeovers in the Toyota production system. Organ Sci 10:43–68

Adner R, Levinthal DA (2004) What is not a real option: considering boundaries for the application of real options to business strategy. Acad Manag Rev 29(1):74–85

Ahuja MJ, Carley KM (1999) Network structure in virtual organizations. Organ Sci 10(6):741–757

Alavi M, Leidner DE (2001) Knowledge management and knowledge management systems: conceptual foundations and research issues. MIS Quarterly 25:107–136

Andriopoulos C, Lewis MW (2009) Exploitation-exploration tensions and organizational ambidexterity: managing paradoxes of innovation. Organ Sci 20(4):696–717

Armstrong CP, Sambamurthy V (1999) Information technology assimilation in firms: the influence of senior leadership and IT structures. Inform Syst Res 10(4):304–327

Barnett ML (2008) An attention-based view of real options reasoning. Acad Manag Rev 33(3):606–628

Bartlett CA, Ghoshal S (1988) Organizing for worldwide effectiveness: the transnational solution. Calif Manag Rev 27(3):54–74

Baum JAC, Calabrese T, Silverman BS (2000) Don't go it alone: alliance network composition and startups' performance in Canadian biotechnology. Strategic Manag J March Special Issue 21:267–294

Benner MJ, Tushman ML (2003) Exploitation, exploration, and process management: the productivity dilemma revisited. Acad Manag Rev 28(2):238–256

Bouzdine-Chameeva T, Dupouët O (2008) Balancing exploration and exploitation: a formal comparison of punctuated equilibrium and ambidexterity, cahier CEREBEM, no 124–08

Bower JL, Christensen CM (1995) Disruptive technologies: catching the wave. Harvard Business Rev 73(1):43–53

Bowman EH, Hurry D (1993) Strategy through the option lens: an integrated view of resource investments and the incremental choice process. Acad Manag Rev 18(4):760–782

Brion S, Favre-Bonté V, Mothe C (2008) Quelles formes d'ambidextrie pour combinerinnovations d'exploitation et d'exploration? Manag Intern 12(3): 29–44

Brown SL, Eisenhardt KM (1997) The art of continuous change: linking complexity theory and time-paced evolution in relentlessly shifting organizations. Adm Sci Q 42:1–34

Burgelman RA (1991) Intraorganizational ecology of strategy making and organizational adaptation: theory and field research. Organ Sci 2:239–262

Burgelman RA (2002) Strategy as vector and the inertia of co-evolutionary lock-in. Admin Sci Quart 47:325–357

Burns LR, Wholey DR (1993) Adoption and abandonment of matrix management programs: effects of organizational characteristics and interorganizational networks. Acad Manag J 36(1):106–138

Cardinal LB (2001) Technological innovation in the pharmaceutical industry: the use of organizational control in research and development. Organ Sci 12:19–36

Childs PD, Triantis AJ (1999) Dynamic R&D investment policies. Manage Sci 45(10): 1359–1377

Christensen M, Knudsen T (2010) Design of decision-making organizations. Manage Sci 56(1): 71–89

Cowan R, Jonard N, Zimmermann J-B (2007) Bilateral collaboration and the emergence of innovation networks. Manag Sci 53(7):1051–1067

Davis SM, Lawrence PR (1977) Matrix. Reading, Mass, Wesley, Boston

De Sanctis G, Monge P (1999) Introduction to the special issue: communication processes for virtual organizations. Organ Sci 10(6):693–703

Dutton JE, Ashford SJ (1993) Selling issues to top management. Acad Manag Rev 18:397–428

Fang C, Lee J, Schilling MA (2010) Balancing exploration and exploitation through structural design: the isolation of subgroups and organizational learning. Organ Sci 21(3):625–642

Farjoun M (2010) Beyond dualism: stability and change as a duality. Acad Manag Rev 35(2):202–225

Ferris GR, Arthur M, Berkson HM, Kaplan DM, Harrell-Cook G, Frink DD (1998) Toward a social context theory of the human resource management–organizational effectiveness relationship. Hum Res Manag Rev 8:235–264

Ford JD, Ford LW (1994) Logics of identity, contradiction, and attraction in change. Acad Manag Rev 19:756–785

Galbraith JR (1973) Designing complex organizations. Reading, Wesley, Mass, Reading

Garud R, Nayyar PR (1994) Transformative capacity: continual structuring by intertemporal technology transfer. Strateg Manag J 15:365–385

Gavetti G, Levinthal D (2000) Looking forward and look backward: Cognitive and experiential search. Admin Sci Quart 45:113–137

Gersick CG (1991) Revolutionary change theories: a multilevel exploration of the punctuated equilibrium paradigm. Acad Manag Rev 16:10–36

Ghoshal S, Bartlett CA (1994) Linking organizational context and managerial action: The dimensions of quality of management. Strategic Manag J 15:91–112

Gibson C, Birkinshaw J (2004) The antecedents, consequences, and mediating role of organizational ambidexterity. Acad Manag J 47:209–226

Gilbert CG (2005) Change in the presence of residual fit: can competing frames co-exist? Organ Sci 17(1):150–167

Gilbert CG (2006) Change in the Presence of Residual Fit: Can Competing Frames Co-exist? Organ Sci 17:150–167

Gupta AK, Smith KG, Shalley CE (2006) The interplay between exploration and exploitation. Acad Manag J 49(4):693–706

Gupta AK, Govindarajan V (2000) Knowledge flows within multinational corporations. Strateg Manag J 21:473–496

Gupta AK, Govindarajan V (1991) Knowledge flows and the structure of control within multinational corporations. Acad Manag Rev 16(4):768–792

Haas M, Hansen M (2007) Different knowledge, different benefits: toward a productivity perspective on knowledge sharing in organizations. Strateg Manag J 28(11):p1133–p1153

Hansen MT (1999) The search-transfer problem: the role of weak ties in sharing knowledge across organization subunits. Admin Sci Quart 44:82–111

Hansen MT, Nohria N, Tierney T (1999) What's your strategy for managing knowledge? Harv Bus Rev 77(2):106–115

He ZL, Wong PK (2004) Exploration versus exploitation: an empirical test of the ambidexterity hypothesis. Organ Sci 15:481–494

Hill CWL, Rothaermel FT (2003) The performance of incumbent firms in the face of radical technological innovation. Acad Manag Rev 28(2):257–274

Huselid MA (1995) The impact of human resource management practices on turnover, productivity, and corporate financial performance. Acad Manag J 38:635–672

Ishida K, Ohta T (2001) On a mathematical comparison between hierarchy and network with a classification of coordination structures. Comput Math Org Theory 7:311–330

Jansen JJP, Tempelaar MP, Van Den Bosch FAJ, Volberda HW (2009) Structural differentiation and ambidexterity: the mediating role of integration mechanisms. Organ Sci 20(4):797–811

Jansen JJP, Van Den Bosch FAJ, Volberda HW (2006) Exploratory innovation, exploitative innovation, and performance: effects of organizational antecedents and environmental moderators. Manag Sci 52(11):1661–1674

Jensen KW, Håkonsson DD, Burton RM, Obel B (2009) Embedding virtuality into organization design theory: virtuality and its information processing consequences. In: Bøllingtoft A, Håkonsson DD, Nielsen JF, Snow CC, Ulhøi J (eds.) New approaches to organization design: theory and practice of adaptive enterprises information and organization design series, vol 8. Springer Publisher, Boston

Joyce WF (1986) Matrix organization: a social experiment. Acad Manag J 29(3):536–561

Kane GC, Alavi M (2007) Information technology and organizational learning: an investigation of exploration and exploitation processes. Organ Sci 18(5):796–812

Kang S-C, Morris SS, Snell SA (2007) Relational archetypes, organizational learning, and value creation: extending the human resource architecture. Acad Manag Rev 32(1):236–256

Kang S-C, Snell SA (2009) Intellectual capital architectures and ambidextrous learning: a framework for human resource management. J Manag Studies 46(1):65–92

Kang S-C, Snell SA (2010) Intellectual capital architectures and ambidextrous learning: a framework for human resource management. J Manag Stud 46(1):65–92

Katila R, Ahuja G (2002) Something old, something new: A longitudinal study of search behavior and new product introduction. Acad Manag J 45(6):1183–1194

Katz R, Allen TJ (1985) Project performance and the locus of performance in the R&D matrix. Acad Manag J 28(1):67–87

King WR, Sethi V (1999) An empirical assessment of the organization of transnational information systems. J Manag Inform Syst 15(4):7–28

Lahiri N (2010) Geographic distribution of R&D activity: how does it affect innovation quality? Acad Manag J 53(5):1194–1209

Lakshman C (2008) Knowledge leadership: tools for top executives. Sage Response Books, New Delhi, India

Lakshman C, Parente R (2008) Supplier-focused knowledge management in the automobile industry and its implications for product performance. J Manage Stud 45:317–342

Lane PJ, Koka BR, Pathak S (2006) The reification of absorptive capacity: a critical review and rejuvenation of the construct. Acad Manag Rev 31(4):833–863

Lawrence P, Lorsch J (1967) Differentiation and integration in complex organizations. Adm Sci Q 12:1–30

Lepak DP, Snell SA (1999) The human resource architecture: toward a theory of human capital allocation and development. Acad Manag Rev 24:31–48

Levinthal DA, March JG (1993) The myopia of learning. Strategic Manag J 14:95–112

Li Y, Vanhaverbeke W, Schoenmakers W (2008) Exploration and exploitation in innovation: reframing the interpretation. Creativity Innov Manag 17(2):107–126

Lindic J, Baloh P, Ribiere VM, Desouza KC (2011) Deploying information technologies for organizational innovation: Lessons from case studies. Intern J Information Manag 31:183–188

MacDuffie JP (1995) Human resource bundles and manufacturing performance: organizational logic and flexible production systems in the world auto industry. Ind Labor Relat Rev 48:197–221

McGrath RG (2001) Exploratory learning, innovative capacity, and managerial oversight. Acad Manag J 44:118–131

Magnusson MG (2004) Managing the knowledge landscape of an MNC: knowledge networking at Ericsson. Knowl Proc Manag 11(4):261–272

March JG (1991) Exploration and exploitation in organizational learning. Organ Sci 2:71–87

March JG (1996) Continuity and change in theories of organizational action. Admin Sci Quart 41:278–287

March JG (2006) Rationality, foolishness, and adaptive intelligence. Strategic Manag J 27:201–214

Maritan CA (2001) Capital investment as investing in organizational capabilities: an empirically grounded process model. Acad Manag J 44:513–531

Martin JA, Eisenhardt KM (2010) Rewiring: Cross-Business-Unit Collaborations in Multi-business Organizations. Acad Manag J 53:265–301

Miller KD, Zhao M, Calantone RJ (2006) Adding interpersonal learning and tacit knowledge to March's exploration-exploitation model. Acad Manag J 49:709–722

Miles MB, Huberman AM (2003). Analyse des données qualitatives. 2nd ed. (trad. 1994). De Boeck

Mirow C, Hoelzle K, Gemuenden HG (2007) The ambidextrous organization in practice: barriers to innovation within research and development. Academy of Management Proceedings

Mom TJM, Van Den Bosch FAJ, Volberda HW (2007) Investigating managers' exploration and exploitation activities: the influence of top-down, bottom-up, and horizontal knowledge inflows. J Manag Stud 44(6):910–931

Mom TJM, Van Den Bosch FAJ, Volberda HW (2009) Understanding variation in manager's ambidexterity: investigating direct and interaction effects of formal structural and personal coordination mechanisms. Organ Sci 20(4):812–828

Mowery DC, Oxley JE, Silveran BS (1998) Technological overlap and interfirm cooperation: implications for the resource-based view of the firm. Res policy 27:507–523

Nahapiet J, Ghoshal S (1998) Social capital, intellectual capital, and the organizational advantage. Acad Manag Rev 23(2):242–266

Nerkar A (2003) Old is gold? The value of temporal exploration in the creation of new knowledge. Manag Sci 49:211–229

Nickerson JA, Zenger TR (2002) Being efficiently fickle: a dynamic theory of organizational choice. Organ Sci 13(5):547–566

Nonaka I, Takeuchi H (1995) The knowledge creating company. Oxford University Press, New York

Nooteboom B (2000) Learning and innovation in organization and economics. Oxford University Press, Oxford

O'Reilly CA, Tushman ML (2004) The ambidextrous organization. Harv Bus Rev 82(4):74–81

O'Reilly CA, Tushman ML (2008) Ambidexterity as a Dynamic Capability: Resolving the Innovator's Dilemma. Res Organ Behavior 28:185–206

Ohja A, Brown JL, Phillips N (1997) Change and revolutionary change: formalizing and extending the punctuated equilibrium paradigm. Comput Math Org Theory 3(2):91–111

Ouchi WG (1980) Markets, bureaucracies, and clans. Adm Sci Q 25:129–141

Peretto P, Smulders S (2002) Technological distance, growth and social effects. Econ J 112:603–624

Pfeffer J (1981) Power in organizations. Pitman Publishing, Mass, Marshfield

Raisch S (2008) Balanced structures: designing organizations for profitable growth. Long Range Plan 41:483–508

Raisch S, Birkinshaw J (2008) Organizational Ambidexterity: Antecedents, Outcomes, and Moderators. J Manag 34:375–409

Raisch S, Birkinshaw J, Probst G, Tushman ML (2009) Organizational ambidexterity: balancing exploitation and exploration for sustained performance. Org Sci 20(4):685–695

Reagans R, McEvily B (2003) Network structure and knowledge transfer: the effects of cohesion and range. Adm Sci Q 48(2):240–267

Rogers M (2003) It's the apple of his eye, Newsweek, 1 Mar 2003

Rosenkopf L, Nerkar A (2001) Beyond local search: Boundary-spanning, exploration, and impact in the optical disk industry. Strategic Manag J 22:287–306

Sabherwal R, Becerra-Fernandez I (2003) An empirical study of the effect of knowledge management processes at individual, group, and organizational levels. Decis Sci 34(2):225–260

Sastry AM (1997) Problems and Paradoxes in a Model of Punctuated Organizational Change. Admin Sci Quart 42(2):237–275

Sanchez R, Mahoney JT (1997) Modularity, flexibility, and knowledge management in product and organization design. IEEE Eng Manage Rev 25(4):50–61

Schreyögg G, Kliesch-Eberl M (2007) How dynamic can organizational capabilities be? Towards a dual process model of capability dynamisation. Strateg Manag J 28:913–933

Schreyögg G, Sydow J (2010) Organizing for Fluidity? Dilemmas of New Organizational Forms. Organ Sci 21:1251–1262

Schulz M (2001) The uncertain relevance of newness: organizational learning and knowledge flows. Acad Manag J 44(4):661–681

Sidhu JS, Commandeur HR, Volberda HW (2007) The Multifaceted Nature of Exploration and Exploitation: Value of Supply, Demand, and Spatial Search for Innovation. Organ Sci 18:20–38

Siggelkow N (2002) Evolution toward Fit. Admin Sci Quart 47:125–159

Siggelkow N, Levinthal DA (2003) Temporarily divide to conquer: centralized, decentralized, and reintegrated organizational approaches to exploration and adaptation. Org Sci 14(6):650–669

Siggelkow N, Rivkin JW (2005) Speed and search: designing organizations for turbulence and complexity. Organ Sci 16:101–122

Siggelkow N (2007) Persuasion with Case Studies. Acad Manag J 50:20–24

Simsek Z, Heavey C, Veiga JF, Souder D (2009) A Typology for Aligning Organizational Ambidexterity's Conceptualizations, Antecedents, and Outcomes. J Manag Studies 46:865–894

Stieglitz N, Heine K (2007) Innovations and the role of complementarities in a strategic theory of the firm. Strateg Manag J 28(1):1–15

Sutton RS, Barto AG (1998) Reinforcement learning: an introduction. MIT Press, Cambridge, MA

Taylor A, Greve H (2006) Superman or the fantastic four? knowledge combination and experience in innovative teams. Acad Manag J 49:723–740

Taylor A, Helfat CE (2009) Organizational linkages for surviving technological change. Organ Sci 20(4):718–739

Teigland R, Wasko MM (2003) Integrating knowledge through information trading: examining the relationship between boundary spanning communication and individual performance. Decis Sci 34(2):261–286

Tortoriello M (2008) Getting the most out of your network: social structure, formal boundaries and knowledge activation, Proceedings of the academy of management meetings, Anaheim

Tran Y, Mahnke V, Ambos B (2010) The effect of quantity, quality and timing of headquarters-initiated knowledge flows on subsidiary performance. Mang Int Rev 50:493–511

Turner KL, Makhija MV (2006) The role of organizational controls in managing knowledge. Acad Manag Rev 31(1):197–217

Tushman ML, Romanelli E (1985) Organizational revolution: a metamorphosis model of convergence and reorientation. Res Organ Behavior 7:171–222

Tushman ML, O'Reilly CA (1996) Ambidextrous organizations: managing evolutionary and revolutionary change. Calif Manag Rev 38(4):8–29

Viitala R (2004) Towards Knowledge Leadership. Leadersh Org Dev J 25(5/6):528–544

Weick KE, Quinn RE (1999) Organizational change and development. Ann Rev Psychology 50:361–386

Westerman GF, McFarlan W, Iansiti M (2006) Organization design and effectiveness over the innovation life cycle. Organ Sci 17(2):230–238

Wright PM, Snell SA, Dunford B (2001) Human resources and the resource based view of the firm. J Manag 27:701–721

Yin RK (1994) Case Study Research, Design and Methods, 2nd edn. Sage Publications, Newbury Park

Young GJ, Charns MP, Heeren TC (2004) Product-line management in professional organizations: an empirical test of competing theoretical perspectives. Acad Manag J 47(5):723–734

Zahra SA, George G (2002) Absorptive capacity: a review, reconceptualization, and extension. Acad Manag Rev 27:185–203

Zander U, Kogut B (1995) Knowledge and the speed of the transfer and imitation of organizational capabilities: an empirical test. Organ Sci 6(1):76–92

Zellmer-Bruhn M, Gibson C (2006) Multinational organization context: implicaitons for team leanring and performance. Acad Manag J 49(3):501–518

SpringerBriefs on Digital Spaces

SpringerBriefs on Digital Spaces is an international research program—the ISD—launched in 2009 by the CIGREF Foundation (www.fondation-cigref.org). The series aims at making a set of concepts, ideas and results of projects carried out under the program available to the research, business and policy communities. ISD—Information Systems Dynamics, is a research program of public interest that works to evaluate the societal and managerial challenges related to the long-term use of information systems and digitality.

Since its launch in 2009, the program has already supported more than 30 projects conducted by international teams from different academic backgrounds (Computer Science, Management Science, Economics, Sociology, Geography and Anthropology) as well as from different geographical regions (Europe, North America and Asia).

The program works on the premise that the *spatial dimension* of the use of digital systems and artefacts is a critical perspective for understanding the dynamics of value creation—and more generally of socio-economizing—in our economies and societies. Understanding emerging practices in digital spaces is a key step toward delineating and conceptualizing a substantial part of the emerging paradigms of economic activities in the twenty-first century. *SpringerBriefs in Digital Spaces* publishes research findings and monographs related to the different facets of these issues. By doing so, the series seeks to contribute to the necessary dialogue between the researchers, practitioners and public policymakers involved in these very critical and rapidly changing fields of research and action.

Editor
The series is edited by Ahmed Bounfour, Professor, European Chair on Intellectual Capital Management, University Paris-Sud, and General Rapporteur of the ISD program.

O. Dupouet et al., *Innovation from Information Systems*,
SpringerBriefs in Digital Spaces, DOI: 10.1007/978-3-642-32876-3,
© The Author(s) 2013